International Diffusion of Technology: The Case of Semiconductors

Studies in the Regulation of Economic Activity

TITLES PUBLISHED

The Dilemma of Freight Transport Regulation
Ann F. Friedlaender

Technological Change in Regulated Industries
William M. Capron, Editor

Reforming Regulation: An Evaluation of the Ash Council Proposals
A staff paper by Roger G. Noll

International Diffusion of Technology: The Case of Semiconductors
John E. Tilton

Studies in the Regulation of Economic Activity

International Diffusion of Technology: The Case of Semiconductors

JOHN E. TILTON

The Brookings Institution / Washington, D.C.

THE BROOKINGS INSTITUTION
1775 Massachusetts Avenue, N.W., Washington, D.C. 20036

ISBN 0-8157-8458-9
Library of Congress Catalog Card Number 72-161593

1 2 3 4 5 6 7 8 9

THE BROOKINGS INSTITUTION is an independent organization devoted to nonpartisan research, education, and publication in economics, government, foreign policy, and the social sciences generally. Its principal purposes are to aid in the development of sound public policies and to promote public understanding of issues of national importance.

The Institution was founded on December 8, 1927, to merge the activities of the Institute for Government Research, founded in 1916, the Institute of Economics, founded in 1922, and the Robert Brookings Graduate School of Economics and Government, founded in 1924.

The general administration of the Institution is the responsibility of a Board of Trustees charged with maintaining the independence of the staff and fostering the most favorable conditions for creative research and education. The immediate direction of the policies, program, and staff of the Institution is vested in the President, assisted by an advisory committee of the officers and staff.

In publishing a study, the Institution presents it as a competent treatment of a subject worthy of public consideration. The interpretations and conclusions in such publications are those of the author or authors and do not necessarily reflect the views of the other staff members, officers, or trustees of the Brookings Institution.

Foreword

INTEREST IN THE ECONOMICS of technological change has grown significantly over the past twenty years. Considerable attention has been given, for example, to the effects of market structure and company size on the level and quality of research and development. Numerous studies of still other determinants of innovative performance are now available. But at least one important aspect of technological change—the diffusion of technology from country to country—has been largely overlooked.

Because the development of new technology is highly concentrated in a few countries, the manner in which it is shared throughout the world has important consequences for economic growth, per capita income, comparative advantage, and corporate subsidiaries abroad. This study, conducted by John E. Tilton while he was a Brookings research associate during 1967–70, provides new insights into the diffusion process by examining the dissemination of semiconductor technology in the United States, Great Britain, France, Germany, and Japan.

The author expresses his appreciation to Merton J. Peck for encouraging him to undertake the study and for valuable assistance throughout its course. He is also grateful to Anthony M. Golding, who generously shared the data and ideas he developed in preparing his doctoral dissertation on the British and American semiconductor industries. Numerous employees of semiconductor firms, trade associations, government agencies, and private research organizations in the United States and abroad furnished information essential to the study. The author cannot thank them individually, but he is grateful to William T. Ellis and Charles Geoffrin for making many of these productive associations possible.

Among the persons who read and commented on earlier versions of the manuscript were William M. Capron, Richard N. Cooper, P. D. Henderson, Herbert S. Kleiman, Dennis C. Mueller, Richard R. Nelson, Roger G. Noll, Hugh T. Patrick, Keith Pavitt, William V. Rapp, and F. M. Scherer, all of whom made helpful suggestions. The author also wishes to thank Barbara Fechter, Jae-Wook Kim, and George F. Kopits for research assistance; Evelyn P. Fisher and Aklog Birara for checking the accuracy of data and sources; Elizabeth H. Cross for editing the manuscript; and Joan C. Culver for preparing the index.

This volume is the fourth publication in the series of Brookings Studies in the Regulation of Economic Activity. Its preparation was assisted by grants from the Ford Foundation and the Alfred P. Sloan Foundation. The views and opinions expressed are, of course, those of the author and should not be attributed to the trustees, officers, or staff members of the Ford Foundation, the Sloan Foundation, or the Brookings Institution.

KERMIT GORDON
President

June 1971
Washington, D.C.

Contents

TABLES

FIGURES

CHAPTER ONE

Introduction

A FEW ADVANCED countries, and in particular the United States, create an unusually large share of the world's new technology. From these innovation centers, new products and processes spread throughout the world. At first, international trade may serve as the channel of diffusion, but eventually the new technology is transferred directly via licenses, international firms, or independent development in other countries. The speed and manner whereby the rest of the world acquires new technology have important consequences. They affect the gaps in technology between countries, the ubiquity of foreign subsidiaries, the shifting pattern of comparative advantage in trade, and international differences in per capita income.

Fear that superior American technology will reduce other countries to mere American satellites has haunted a generation of Europeans. During the last decade this fear has been exacerbated by the proliferation of new American subsidiaries in Europe, the mounting concern over the technology gap, and the dominance of American corporations in vital new industries such as computers and semiconductors. Compounding the European situation is the rapid advance of Japanese technology. With the United States well entrenched in the manufacture of the most advanced goods, with Japan overtaking European technology, and with more and more of the underdeveloped countries manufacturing the less complex products, what will Europe produce? How will it trade? What will happen to its economic growth and standard of living?

The Japanese, on the other hand, worry about how they can prevent foreign firms from setting up subsidiaries and dominating key sectors of their economy without impeding the inflow of foreign technology; how they can become important innovators of technology, as well as swift imitators, and thereby reduce their dependence on foreign technology; and how they can sustain their rapid economic growth and favorable balance of trade, particularly when their wage rates are rising rapidly and they face increasing competition from Hong Kong and other countries that still have a low wage scale. And in the United States, many fear that the rise of American subsidiaries abroad and the swift dissemination of new American technology are undermining the country's ability to compete in international trade, thus aggravating balance-of-payments difficulties.

All these important public policy concerns are directly affected by the process of creating and diffusing new technology. Interest in this process has flourished among economists over the last decade, and much has been learned about the economics of technological change.[1] Still, in many areas our knowledge remains limited and inadequate. This is particularly true for the international diffusion process. Why, for example, do international firms effect the transfer of certain innovations, while others are passed from one independent firm to another through licensing agreements? Would restrictions on the international firm impede the dissemination of new technology? How do new firms and easy entry conditions affect diffusion? Or more generally, what market structure characteristics are conducive to swift diffusion? Why do the channels of diffusion and the type of firms responsible for diffusion often vary from country to country? What significance, if any, do such differences have for diffusion? What exactly do we mean by diffusion, what are its various aspects, and how can we measure it? These are all important questions for which as yet we have no definite answers.

This study attempts to shed additional light on such questions and the international diffusion process in general by investigating the diffusion of the transistor, integrated circuit, and other semiconductor technology in the United States, Britain, France, Germany,[2] and Japan. The central hypothesis examined is the following: The diffusion of new technology is accelerated by a market structure that allows new firms to enter an industry and supplant the established industry leaders whenever the latter fail to employ new techniques as quickly as economic conditions warrant. Hence, diffusion tends to occur faster in countries with flexible market structures than in countries where entry barriers are high and company rankings rigid.

The scope of the study is limited in several ways. First, attention is

1. Surveys of the literature in this field are found in Richard R. Nelson, Merton J. Peck, and Edward D. Kalachek, *Technology, Economic Growth, and Public Policy* (Brookings Institution, 1967); and Edwin Mansfield, *The Economics of Technological Change* (Norton, 1968).

2. Throughout this study, Germany denotes the Federal Republic of Germany, or West Germany.

concentrated on the semiconductor industry in the belief that an in-depth case study of one industry can yield more insights than a less detailed investigation embracing a number of industries. The semiconductor industry was selected for several reasons. (a) It is important. American firms alone annually produce more than a billion dollars' worth of transistors, integrated circuits, and other semiconductors. Moreover, these devices form the heart of modern electronic equipment from pocket radios to computers and guided missiles. (b) The industry is new. Since semiconductor production was negligible before World War II, the inquiry can focus on the postwar period and avoid the data and other problems that a longer time span would entail. (c) The industry is a prime example of a rapid-growth research-intensive industry. It is precisely in such industries, which have become the new symbols of national prestige, that the other advanced countries find the gap between American technology and their own the greatest and the most distressing.[3]

Second, the study is confined to diffusion in the United States, Britain, France, Germany, and Japan for several reasons. (a) Although some Communist states and a few of the less developed countries such as Hong Kong and Taiwan are semiconductor producers, these five advanced countries dominate the industry and account for most of the world's production. (b) Statistical data and other information are more readily available for these countries. (c) The process of diffusion in the advanced countries may differ in important respects from that in either the centrally planned economies of the Communist bloc or the underdeveloped countries, and so is best considered separately.

Third, little consideration is given to disparities among countries in cultural and social attitudes or to differences in internal firm organization and managerial efficiency, although such factors are sometimes advanced as important causes of international differences in technological progressiveness. This study finds that the international diffusion of semiconductor technology can be analyzed without recourse to these considerations, which are often difficult to quantify or even define rigorously. This, of course, does not mean that the diffusion of technology is unrelated to attitudes and managerial efficiency, but it does suggest that differences in attitudes and managerial efficiency insofar as they affect diffusion are significantly associated with the market structure characteristics and other factors examined in the study. Thus, public policies that accelerate diffu-

3. See Organisation for Economic Co-operation and Development, *General Report: Gaps in Technology* (Paris: OECD, 1968), p. 17.

sion by influencing market structure may change business attitudes and efficiency as well. For example, a market structure might inhibit the diffusion of technology by stifling competition, fostering a live-and-let-live attitude among businessmen, and allowing firms with inept management to survive. Public policies that alter this structure and promote competition so that diffusion is stimulated would also change attitudes and improve managerial efficiency.

Finally, the emphasis is on diffusion, though the creation of new technology is also considered, for in practice the two are closely intertwined over the technological development cycle through which an innovation must evolve before its full economic impact is realized. For our purposes, this evolution can be separated into the innovative process and the diffusion process.[4]

The innovative process creates new technology. It encompasses such activities as basic research in solid state physics and the other disciplines on which semiconductor technology is based, applied research that leads to the first laboratory model of a new semiconductor device, and the subsequent engineering and development necessary before commercial production is feasible. Similarly, the conception and development of new manufacturing processes and new production equipment are innovative activities.

The first workable model of a new product or process is often referred to as an invention, and its first commercial production or use as an innovation. This usage is followed in this study. However, it is important to note that the innovative process and innovation are not used as synonyms. The former is more comprehensive: it includes activities whose purpose is to develop innovations, as well as other activities, such as basic research, with different objectives.

The diffusion process is responsible for the adoption and exploitation of the new technology generated by the innovative process. It involves primarily production and marketing activities,[5] such as tooling up for production, incorporating new techniques into the manufacturing process, learning to use them efficiently in volume production, and improv-

4. Not all studies make this distinction between the innovative and the diffusion processes. Some instead consider the activities of both as the innovative process. See, for instance, U.S. Department of Commerce, *Technological Innovation: Its Environment and Management* (1967), pp. 8–11.

5. The first commercial production of new semiconductors is an exception. As pointed out above, efforts to solve the basic technical problems encountered in developing semiconductor innovations are considered part of the innovative process.

ing distribution and selling procedures. It is during the diffusion process that the use of a new product or process grows toward an equilibrium level of utilization.[6]

Two characteristics of diffusion, as described here, should be noted. First, diffusion may proceed at several institutional levels; specifically, it may or may not involve the transfer of technology among firms and countries. When one firm incorporates in its manufacturing process new technology developed in another firm's research and development (R&D) facilities (and perhaps used in production by the other firm), diffusion occurs through an interfirm transfer of technology. If these firms are located in different countries, international diffusion of the technology takes place. Diffusion also occurs when a firm introduces into its production operations technology developed by its own R&D laboratory, even though there is no interfirm transfer of technology. Similarly, diffusion takes place when foreign subsidiaries employ technology developed by their parent companies. In this case, international diffusion occurs, but without an interfirm transfer of technology.

Second, in diffusing new technology, firms may be pioneers or imitators. Pioneers in the diffusion process are the first firms to use new technology successfully in production, and thus the first to realize the savings in production costs and the improvements in product performance made possible by the new technology. Pioneering firms need not necessarily be leaders in the innovative process, but they must be quick to recognize the commercial significance of new advances made in their own and other R&D facilities and skilled at adapting them to production. They must be world leaders in improving the production process, thereby reducing fabrication costs and increasing product performance so that what were once high-cost devices with limited applications can reach ever larger markets. In contrast, imitating firms effect diffusion by duplicating the successful production techniques exploited by pioneering firms. Rapid and effective imitation is essential if a firm (or country) is to catch up with the world leaders in using new technology, but to become a leader itself it must on occasion take the initiative and pioneer in the diffusion process.

Both the innovative and diffusion processes are essential for economic progress. A new semiconductor device, no matter how great the advance

6. The diffusion process can be viewed as a movement from an old to a new equilibrium. See Nelson, Peck, and Kalachek, *Technology, Economic Growth, and Public Policy*, pp. 97–98.

The nature of the diffusion process is examined more fully in Chapter 3.

in technology it embodies, makes no contribution until it is produced, marketed, and in the hands of firms and persons who can use it. Conversely, the speed and extent with which an innovation diffuses, and in turn its impact on the industry and economy in general, greatly depend on the magnitude of the advance in technology. Consequently, it is not possible to divorce the innovative process from the diffusion process, and so in assessing the performance of firms and nations in diffusion, their innovative achievements are considered too.

The next chapter briefly describes the semiconductor industry, its products, and its place in the electronics industry as a whole. It also identifies the major semiconductor innovations, nearly all of which were introduced by American firms.

Chapter 3 examines the nature of the diffusion process, measures the diffusion of semiconductor technology, and investigates semiconductor production and trade. It finds that all five countries under review have quickly adopted new semiconductor technology, even though their success in using this technology to promote domestic production and a favorable balance of trade has varied greatly.

Chapters 4, 5, and 6 identify firms and types of firms active in the semiconductor industries of the United States, Europe, and Japan, respectively. They assess company performance in both the innovative and diffusion processes, and find that the type of firm contributing most to the rapid pace of diffusion differs in each of these areas. New firms are the most aggressive diffusers in the American market, American subsidiaries (having eclipsed the large electrical companies in the sixties) in the European market, and large electrical firms in the Japanese market. To explain why the type of firm responsible for rapid diffusion varies from one country to another and the implications of this finding for our hypothesis regarding easy entry conditions and the speed of diffusion, these chapters also examine licensing policies, economies of scale, learning economies, market size, capital requirements, military demand, government R&D funds, antitrust and competition policies, restrictions on foreign subsidiaries, and other features of the semiconductor industries in these countries.

Chapter 7 ties together the various findings of earlier chapters and draws some general conclusions. This chapter along with the concluding sections found at the end of Chapters 3 through 6 provide an abbreviated tour of the book.

CHAPTER TWO

The Semiconductor Industry

SINCE THE semiconductor industry is a branch of the electronics industry, this chapter begins by briefly describing the structure of the electronics industry and the place of the semiconductor industry in that structure. It then examines the various types of semiconductors and other active components, focusing on their principal characteristics and technological evolution. The final section identifies the major semiconductor innovations and describes their significance.

The Electronics Industry

The origin of the electronics industry is associated with the development of the radio, and until World War II radios and related equipment accounted for nearly all of the industry's output. Since 1939, however, electronic technology has advanced rapidly. Many new products have been developed, and the industry has enjoyed remarkable growth. By 1968, the American market alone was producing $24 billion worth of electronic goods annually, a seventyfold increase over its 1939 sales.[1]

As Figure 2-1 illustrates, the industry produces both final goods and components. Final goods are classified according to end-use as consumer, industrial, and government products; for example, radios and television sets are consumer products, computers and X-ray equipment are industrial products, and radar and missile-guidance systems are government products. At times different end-users need the same or similar electronic products; thus some goods are found in two or even all three sectors. A military radio, for instance, belongs in the government sector, a taxicab dispatch radio in the industrial sector, and an entertainment radio in the consumer sector. Similarly, computers are found in both the industrial and government markets.

1. *Electronic Industries Yearbook, 1969* (Washington: Electronic Industries Association, 1969), p. 2.

Figure 2-1. Sectors of the Electronics Industry

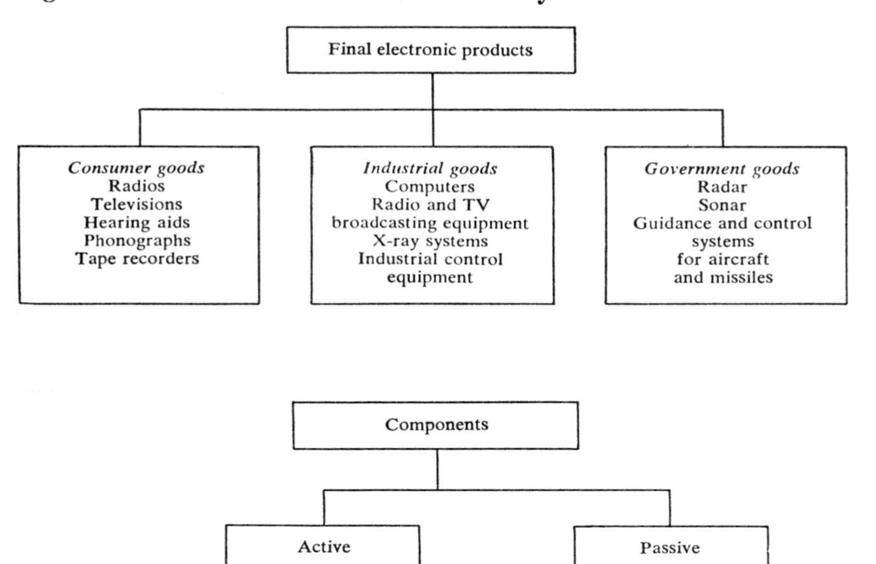

The government sector, which is composed mainly of military equipment, has by far the strictest performance standards in terms of capabilities and reliability. In the consumer sector, costs are a major concern and much lower performance specifications are imposed. The industrial sector falls between the government and consumer sectors in both performance and cost requirements.

The component sectors of the electronics industry produce only certain parts used in final electronic equipment: (1) active devices, which include electron tubes and discrete semiconductors, (2) passive devices such as capacitors, resistors, inductors, and relays, and (3) integrated circuits, which combine active and passive elements in one device. Cabinets, mounts, and many other hardware pieces are inputs from other industries.

Active components affect or modify an electrical signal through amplification, modulation, generation, and switching operations, and are the distinguishing characteristic of electronic (as opposed to other electrical) equipment. Passive components perform different functions; for example, resistors impede the flow of electricity and capacitors store electricity.

Electron Tubes and Semiconductors

Active components have evolved through three technological generations: electron tubes, transistors and other discrete semiconductors, and integrated circuits.[2] This section briefly examines the development of each generation and identifies the types of devices in each.

Electron Tubes

Electron tubes, or vacuum tubes and valves as they are sometimes called, were invented early in this century. Commercial production began in 1920. Research and development (R&D) have since produced a variety of different types and models that are generally grouped into three subclasses—receiving, cathode ray, and special purpose tubes. The receiving tube is used to detect and amplify an electrical signal in radios, television sets, tape recorders, and other equipment. By far the most common cathode ray tube is the television picture tube. Special purpose tubes comprise many different types, but most are designed for high frequency or power applications.

For many years, the electron tube was the only significant active component available, and the development of new final electronic goods depended on its technical capabilities. Consequently, a tremendous effort was undertaken to improve and perfect the electron tube. In certain

2. Some regard the Hertz oscillator, the spark gap, and the coherer as the first generation of active components and electron tubes as the second. However, the common usage of these terms is that given in the text. See P. E. Haggerty, "Integrated Electronics and Change in the Electronics Industry" (speech delivered to the International Electron Devices Meeting, Oct. 19, 1967; processed).

cases, such as that of the color television picture tube, the effort continues even today.

As their capabilities increased, important and inherent limitations of electron tubes became clear. First, except for power tubes, they are inefficient. To operate, they must be heated and energized. Heating uses power and frequently requires that other components be protected from heat damage. Second, tubes have limited reliability; they burn out and must be replaced. This creates a major problem in equipment containing thousands of tubes such as earlier telephone exchange systems. Third, the design and composition of tubes inhibit reductions in size and cost. Finally, tubes are fragile and easily damaged by shock. These characteristics, particularly limited efficiency and reliability, encouraged the search for an alternative to the electron tube, which began before World War II and culminated in the development of the transistor after the war.

Discrete Semiconductors

An electron tube alters an electric current as the current passes through the space enclosed by the tube. The enclosed space is either a vacuum or contains an inert gas. By contrast, in both discrete semiconductors and integrated circuits, current alteration depends on the structure and conductivity of solid state materials.

Today, germanium and silicon crystals are the most commonly used semiconductor materials. Hundreds of other semiconductor elements and compounds exist, but only a few, such as gallium arsenide, are used to even a limited extent in semiconductor production. As the name suggests, semiconductor materials conduct electricity better than insulators but not as well as highly conductive metals such as aluminum or copper. Under proper conditions, semiconductors display asymmetrical conduction properties, and their electrical resistance can be varied by heat and light. These properties permit them to rectify, amplify, and perform numerous other functions.

Semiconductor production was negligible before World War II. Some semiconductors, particularly galena crystals, had been used in early radios, but were largely displaced after 1925 by electron tubes. During the war, crystal diodes for radar equipment were produced in great numbers. Semiconductor research was stepped up, and continued after the war. In 1948 the invention of the transistor by Bell Laboratories was announced. Three years later Western Electric, an affiliate of Bell Laboratories, began commercial production.[3] These developments stimulated interest in semiconductor technology and induced substantial R&D expenditures throughout the world that produced a multitude of new and improved devices.

The three major kinds of discrete semiconductors, as Figure 2-1 indicates, are transistors, diodes (including rectifiers), and special devices. Within the transistor class is a variety of different types, each designed with special characteristics for specific applications. Although similar in certain respects, one cannot normally be substituted for the other. A high frequency transistor, for instance, cannot replace a power transistor. The same is true of the diode class. A rectifier, which is simply a power diode, cannot be used in a pocket radio powered with batteries. The special category is even more heterogeneous. It includes light-sensitive devices, such as solar cells that convert light into electricity, and special types of transistors and diodes, such as silicon controlled rectifiers and tunnel diodes. Special semiconductors are not produced in large numbers and the value of their output is small compared to that of transistors or diodes. They do, however, include some technically sophisticated items.

Although transistors and diodes perform different functions,[4] their production is similar in many respects, and most transistor firms also manufacture diodes. As transistor production usually requires a more sophisticated technology,[5] the reverse is less common, but even so the proportion of diode firms producing transistors is substantial.

3. In the patent search for the transistor, Western Electric uncovered a number of patents for roughly similar devices. One of the more plausible suggestions was described in a patent awarded in 1930 to Julius Lilienfeld, then a professor of physics at the University of Leipzig. However, none of these early proposals apparently works; see "The Improbable Years," in *The Transistor: Two Decades of Progress, Electronics,* Vol. 41 (Feb. 19, 1968), p. 82.

For a description of the invention and development of the transistor, see Richard R. Nelson, "The Link Between Science and Invention: The Case of the Transistor," in *The Rate and Direction of Inventive Activity: Economic and Social Factors,* A Conference of the Universities–National Bureau Committee for Economic Research and the Committee on Economic Growth of the Social Science Research Council (Princeton University Press for the National Bureau of Economic Research, 1962), pp. 549-83. This study also briefly describes some of the important research in the semiconductor field conducted before World War II.

4. Transistors are commonly used to amplify and switch, and diodes to rectify. Rarely can one be substituted for the other, though transistors can rectify and special diodes can perform many transistor functions. See J. H. Forster and R. M. Ryder, "Diodes Can Do Almost Anything," *Bell Laboratories Record,* Vol. 39 (January 1961), pp. 3-9.

5. Semiconductor materials such as silicon and germanium are doped with

The technology used to produce discrete semiconductors has improved rapidly over the last two decades, and the market for these devices has expanded greatly as a result. When the transistor was first introduced, it was expensive compared to the receiving tube, the one electron tube it could replace in certain applications. Consequently, the transistor was used only where its unique properties were highly desirable. For example, hearing aid producers, attracted by the transistor's low power consumption and small size, were among the first to substitute transistors for receiving tubes. Technological progress rapidly improved reliability, reduced costs, and expanded the potential market into high frequency applications (see Figure 2-2). Research also pushed back the power, temperature, and radiation frontiers and thus opened up additional markets.

These developments have greatly stimulated the use of transistors. Radios, computers, tape recorders, television sets, and many other electronic products have switched to the transistor as its price and performance properties have improved compared to those of the receiving tube.[6] Nor has rapid progress been confined to transistors. Diodes and many special semiconductor devices have enjoyed similar improvements. Such advances have increased the frequency and power capabilities of discrete semiconductors and helped them capture markets from the special purpose tubes. Today only cathode ray tubes as a group remain immune to competition from discrete semiconductors.

Besides stimulating the substitution of semiconductors for tubes, rapid advances in semiconductor capabilities, reliability, and costs have permitted the development of many new electronic products. Portable radios

---

elements either from the fifth group of the periodic table, such as arsenic and phosphorus, or from the third group, such as gallium and boron. They then contain either more or less electrons than called for by their crystal structure. An *n*-type area has an excess of electrons (more negative charges) and a *p*-type area has a deficit of electrons (positive "holes" or charges). The union of an *n*-type area with a *p*-type area constitutes a junction.

Transistors normally have two junctions and three terminals (connections with the rest of the circuitry), while diodes generally have only one junction and two terminals. There are some exceptions to this distinction between transistors and diodes; for example, the silicon controlled rectifier has three junctions, and the MOS (metal oxide semiconductor) transistor only one.

6. Though competition from the transistor has been stiff, receiving tube production is still far from negligible. In 1968 the number of receiving tubes produced in the United States was only 37 percent below the record levels of the mid-1950s.

Figure 2-2. Transistor Performance and Costs in the United States, 1951–65

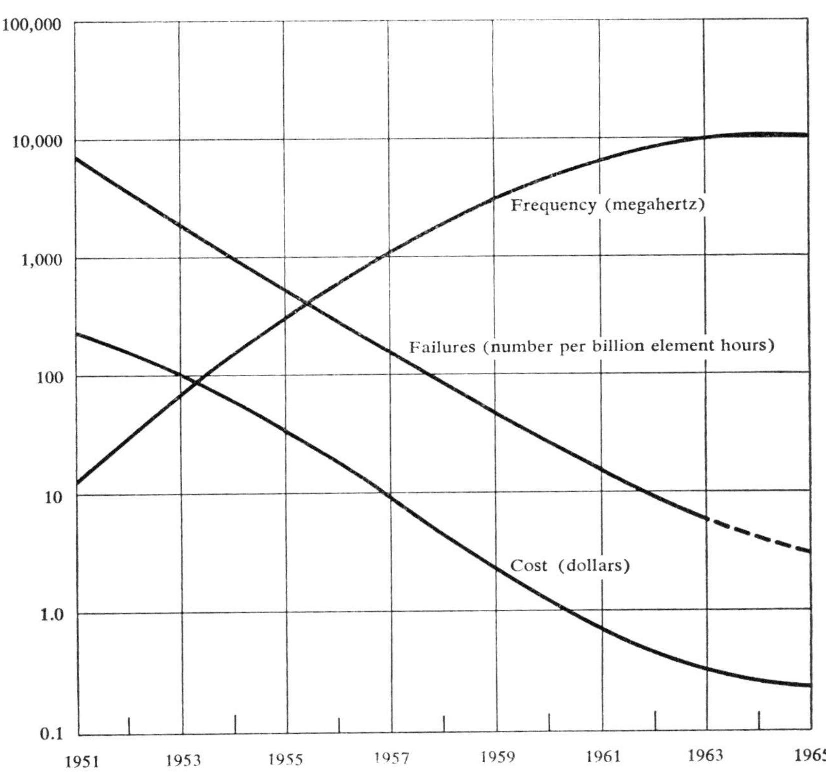

Source: W. C. Hittinger and M. Sparks, "Integrated Circuits," *Bell Laboratories Record,* Vol. 44 (October–November 1966), p. 295. Copyright 1966 Bell Telephone Laboratories, Inc. Used by permission.

that fit into a pocket and computers with much greater speed and capacity than their predecessors are examples.

As the frontiers of semiconductor technology expanded during the fifties, certain intrinsic limitations of discrete semiconductor devices became increasingly apparent, just as the limitations of electron tubes had become apparent a generation earlier. In particular, semiconductors individually packaged, tested, and placed into circuits posed reliability and cost problems that inhibited the development of very large electronic systems, such as telephone exchanges composed of hundreds of thousands

—even millions—of electronic components. Research efforts directed toward overcoming this "tyranny of numbers" led to the invention of the monolithic integrated circuit by Texas Instruments in 1958. Commercial production began three years later with Texas Instruments and Fairchild Camera and Instruments leading the way.

Integrated Circuits

An integrated circuit is a combination of two or more elements (specifically, two or more transistors, diodes, and passive components) which are inseparably associated within a silicon crystal.[7] When this new device was first introduced into the market, it was expensive compared to the discrete semiconductors it could replace and so was used almost exclusively in government equipment, particularly in missiles where performance, not cost, had top priority. Since then, improvements in the capabilities, reliability, and costs of integrated circuits have been as great or greater than the improvements a decade earlier in transistors. Their frequency range has continuously expanded, and the number of elements now contained in some integrated circuits has reached the hundreds, leading some to suggest that such large-scale integration should really be considered as a fourth-generation component. Simultaneously, the average price has fallen precipitously—in the United States prices dropped more than 90 percent between 1963 and 1968.

Although the production of integrated circuits is much more complicated than the production of transistors and diodes, the technologies are similar. Many of the techniques and processes required in manufacturing integrated circuits are also used in the large-batch low-cost fabrication of

7. There are two types of integrated circuits—monolithic and hybrid. The elements of a monolithic integrated circuit are generally contained *in* the substrate, which is a silicon crystal. The transistor and diode elements are produced along with the passive elements in one continuous production process. In contrast, hybrid integrated circuits, which appeared before 1958, have a substrate made from an insulating material, commonly glass or ceramics. The passive elements are created *on* the substrate, and then discrete semiconductor devices or monolithic integrated circuits that have been separately manufactured are affixed.

Hybrid integrated circuits constitute an advance primarily in passive component technology, not in semiconductor technology. They do not provide an alternative device or an alternative technology to perform the active functions of discrete semiconductor devices and monolithic integrated circuits. Indeed, the latter are an integral part of hybrid circuits. For our analysis, then, only monolithic integrated circuits are relevant, and unless otherwise noted, the term integrated circuits refers solely to monolithic integrated circuits.

discrete semiconductors. As a result, nearly all integrated circuit firms also produce transistors and diodes.

Major Semiconductor Innovations

Advances in semiconductor technology since 1948 have caused the prices of transistors, integrated circuits, and other semiconductor devices to fall while reliability and capabilities have improved. In part, this progress can be attributed to many small advances in technology and managerial improvements,[8] but a good share of the credit must go to major innovations. This section, starting with the first transistor, identifies the innovations that produced a significant technical advance and have had, or will have, a major impact on the industry. These innovations include both new semiconductor devices and new processes for manufacturing semiconductors.

The major product innovations are listed in Table 2-1, which also identifies the firm or firms (and their nationality when not American) primarily responsible for the first commercial production, the date commercial production began, and the primary contribution or importance of each innovation.[9] Table 2-2 provides similar information for the major process innovations. In addition, when a new process led directly to one of the new products indicated in Table 2-1, Table 2-2 identifies the product. In such cases, the same firm nearly always accounts for both, and it is probably more appropriate to think of the new process and new product as only one major innovation.

The information contained in Tables 2-1 and 2-2 requires three caveats. First, such lists of major innovations are to some extent arbitrary. The innovations included in the two tables were selected after discussions with scientists, engineers, and market specialists in the semiconductor industry, who, while they occasionally designated different firms and slightly different dates for a few of the innovations, were in general

8. For an illustration of how one plant, the International Telephone and Telegraph Corporation (ITT) diode plant at Lawrence, Mass., reduced production costs with small inexpensive improvements, see "Diodes—They Also Serve," *Electronics,* Vol. 41 (Feb. 19, 1968), p. 90.

9. No attempt is made to describe the technical aspect of these innovations. The interested reader can find this information in lay terminology in trade publications such as *Electronics* or *Electronic News,* and in company periodicals such as the *Bell Laboratories Record.*

Table 2-1. Major Product Innovations in the Semiconductor Industry, 1951–68[a]

Innovation	*Principal firm responsible*	*First commercial production*	*Importance*
Point contact transistor	Western Electric	1951	First solid state amplifier. More efficient in power consumption, and eventually less costly, more reliable, and smaller than tubes.
Grown junction transistor	Western Electric	1951	Increased production yield, thus lowering costs. Less electrical noise and greater resistance to shock.
Alloy junction transistor	General Electric RCA	1952	Greatly improved transistor capability to perform digital (switching) operations. Encouraged development of second-generation computers.
Surface barrier transistor	Philco	1954	Increased transistor frequency range and switching speeds; useful in computer development.
Silicon junction transistor	Texas Instruments	1954	First transistor not made from germanium. Silicon increased temperature range of operation, thus opening up military market. Also increased frequency range.
Diffused transistor	Western Electric Texas Instruments	1956	Lower production costs; increased reliability and frequency range.
Silicon controlled rectifier	General Electric	1956	Valve allowing electric current to flow in one direction only, at same time controlling the flow. Can replace thyratron tubes for control and switching functions.
Tunnel diode[b]	Sony (Japan)	1957	Can replace special purpose tubes for amplification and oscillation at very high frequencies. Very fast, but so far too expensive: though a major technical development, commercial use is limited.
Planar transistor	Fairchild	1960	Batch production possible, lowering costs. Improved performance and reliability.
Epitaxial transistor	Western Electric	1960	Increased switching speed; lower production costs.
Integrated circuit	Texas Instruments Fairchild	1961	First semiconductor device with two or more elements within a silicon substrate. Incorporated bigger segment of circuit into one device, making increased reliability, faster switching speeds, lower costs, and greater miniaturization feasible.
MOS transistor	Fairchild	1962	Cheaper slow-speed switch. Easy to integrate into circuit designs. Fewer steps in production process.
Gunn diode[b]	International Business Machines	1963	Gallium arsenide device, can replace klystron and magnetron tubes for generation and oscillation in microwave range. Still in experimental and development stage.

Sources: See text, pp. 15 and 18.

a. From 1963 to 1968, important advances in semiconductor technology were concentrated in the integrated circuit field. These innovations are considered further developments of integrated circuit technology and are not separately identified here. A list is given in Anthony M. Golding, "The Semiconductor Industry in Britain and the United States: A Case Study in Innovation, Growth and the Diffusion of Technology" (Ph.D. dissertation, University of Sussex, forthcoming).

b. Company and date indicated are for the first laboratory model rather than the first commercial production.

Table 2-2. Major Process Innovations in the Semiconductor Industry, 1950–68[a]

Innovation	*Principal firm responsible*	*Date of development*	*Associated product innovation*[b]	*Importance*
Single crystal growing	Western Electric	1950	Grown junction transistor	Method of growing and doping germanium crystals. Bell Laboratories (an affiliate of Western Electric) achieved same innovation for silicon crystals in 1952, leading to silicon junction transistor.
Zone refining	Western Electric	1950		Produced extremely pure germanium and silicon crystals. Also improved doping process.
Alloy process	General Electric	1952	Alloy junction transistor	New method for forming junctions, leading to transistors with superior switching capabilities.
3–5 compounds	Siemens (Germany)	1952		Semiconductor materials made from combinations of elements in third and fifth groups of periodic table, such as gallium arsenide. Later used in the Gunn diode.
Jet etching	Philco	1953	Surface barrier transistor	Process for producing transistors with increased frequency and switching properties.
Oxide masking and diffusion[c]	Western Electric	1955	Diffused transistor	Improved method for forming junctions. Batch production possible, reducing production costs. Also improved quality control; increased power and frequency capabilities of transistors, diodes, and rectifiers.
Planar process	Fairchild	1960	Planar transistor	Development on oxide masking and diffusion process that lowered production costs and improved performance characteristics; of great importance for economical production of integrated circuits.
Epitaxial process	Western Electric	1960	Epitaxial transistor	Technique for junction forming whereby one type of crystal structure is grown on another. Used with planar process, it reduces production costs and increases performance characteristics, particularly frequency range, of transistors and integrated circuits.
Plastic encapsulation	General Electric	1963[d]		Inexpensive method of protecting transistors and integrated circuits from contamination when reliability is not crucial. Though important commercially, not a major technical advance.
Beam lead	Western Electric	1964		Reduces encapsulation costs for highly reliable semiconductor devices. Permits air isolation of integrated circuit elements, and facilitates mixing of semiconductor and thin-film technologies in hybrid integrated circuits.

Sources: See text, pp. 15 and 18.

a. From 1964 to 1968, important advances in semiconductor technology were concentrated in the integrated circuit field. These innovations are considered further developments of integrated circuit technology and are not separately identified here. A list is given in Golding, "The Semiconductor Industry in Britain and the United States."

b. When the new process led directly to one of the new semiconductor products listed in Table 2-1, this column indicates the product.

c. Up to this point, diffusion has referred to the transfer or dissemination of technology. The term is also used in this study, as it is here, to identify a specific process used in semiconductor production. The meaning intended is apparent from the context.

d. Plastic encapsulation was known in the 1950s but was not practical for commercial use.

agreement on the major innovations. Also encouraging is the similarity between these lists and others found elsewhere.[10] Nevertheless, some questioned including plastic encapsulation and the tunnel diode on the list of major innovations, while others would have added the germanium power rectifier first produced by Associated Electrical Industries (a British firm), the silicon power rectifier first produced by General Electric (an American firm), and the alloy diffused transistor first produced by Philips (a Dutch firm).

Second, the economic and technical importance of the major semiconductor innovations varies greatly.

Third, as Chapter 1 pointed out, a technological advance proceeds through various stages (basic research, applied research, development, commercial introduction, and diffusion) of the technological development cycle before its impact on costs and performance is complete. Tables 2-1 and 2-2 identify only the firms responsible for the first commercial production or use of an innovation. In some cases, different firms performed the early R&D that led eventually to the innovation's successful introduction into the market. Similarly, the firms noted were not always the most successful in marketing and diffusing the innovation they introduced.

Even with due allowance for these considerations, Tables 2-1 and 2-2 indicate the dominance of American firms during the postwar period in introducing the major semiconductor innovations. Among the American companies, Western Electric, building on the research of Bell Laboratories, has been the leader, although in recent years the share of other American companies has been growing.

10. See, for example, U.S. Department of Commerce, Office of Technical Services, *Patterns and Problems of Technical Innovation in American Industry,* Report to the National Science Foundation by Arthur D. Little, Inc. (September 1963), p. 150; Christopher Freeman, "Research and Development in Electronic Capital Goods," *National Institute Economic Review,* No. 34 (November 1965), p. 64; Organisation for Economic Co-operation and Development, *Electronic Components: Gaps in Technology* (Paris: OECD, 1968), p. 44; and Anthony M. Golding, "The Semiconductor Industry in Britain and the United States: A Case Study in Innovation, Growth and the Diffusion of Technology" (Ph.D. dissertation, University of Sussex, forthcoming).

CHAPTER THREE

The Rate of Diffusion

AFTER EXAMINING the general nature of the international diffusion process, this chapter measures the diffusion of semiconductor technology in the United States, the principal innovating country, and in Britain, France, Germany, and Japan. New technology has disseminated rapidly in these countries, apparently as rapidly as warranted by demand conditions. For other reasons, however, the ability of these countries to use this technology to promote domestic production and a favorable balance of trade in semiconductor products has varied greatly.

The Diffusion Process

The scientific and technical resources needed to develop new products and processes and then to carry out essential modifications following their commercial introduction are heavily concentrated in a few advanced countries.[1] These countries also enjoy domestic demand conditions that stimulate a wide range of innovations: high wage rates promote

1. This section draws heavily on recent contributions in international trade theory. Most economists share a growing disenchantment with the traditional factor proportions and neoclassical trade theories. The basically static nature of traditional theories, coupled with accumulating evidence of the important role technological change often plays in determining comparative advantage and trade patterns, has encouraged a number of trade specialists to turn their attention to the diffusion of innovations and technology among countries. See, for example, M. V. Posner, "International Trade and Technical Change," *Oxford Economic Papers,* Vol. 13 (October 1961), pp. 323–41; G. C. Hufbauer, *Synthetic Materials and the Theory of International Trade* (Harvard University Press, 1966); Raymond Vernon, "International Investment and International Trade in the Product Cycle," *Quarterly Journal of Economics,* Vol. 80 (May 1966), pp. 190–207; Harry G. Johnson, *Comparative Cost and Commercial Policy Theory for a Developing World Economy* (Stockholm: Almqvist and Wiksell, 1968); and Raymond Vernon (ed.), *The Technology Factor in International Trade,* A Conference of the Universities–National Bureau Committee for Economic Research (Columbia University Press for the National Bureau of Economic Research, 1970).

labor-saving innovations; high personal incomes generate strong demand for new products from affluent consumers; and large military and space programs support many innovations. In the less developed countries, the evolution of demand tends to follow the pattern in advanced countries as wage rates and income levels rise; hence, innovations induced by the structure of demand at a particular income level are normally first made in more advanced countries. These factors also account in large part for the dominant position of the United States among the advanced countries in innovation, for similar differences exist between the United States and other advanced countries as are found between the advanced countries in general and the less developed countries.

After successful introduction in the innovating country, new products and processes generally diffuse abroad. The first imitators normally are advanced countries, whose technical capabilities and resources permit the early identification of significant foreign developments and facilitate the actual transfer of technology. In addition, because the structure of demand is similar in advanced countries, there is a strong incentive for adopting innovations introduced in other advanced countries. The less developed countries often attract the eventual transfer of new technology by their lower wage rates, but usually the process is slower.

The steps in the international diffusion of a new product (or a product produced with a new process) are depicted in Figure 3-1, which illustrates trends in production, consumption, exports, and imports in the innovating country, an early imitating country, and a late imitating country. Consumption in the innovating country generally occurs as soon as the new product is successfully produced and placed on the market. Later, consumption begins in other countries.[2] This "demand lag" varies from one imitating country to another. In general, the demand lag tends to be shorter in countries in which the size of the economy is larger, the similarity in the structure of demand between the imitating and the innovating countries is greater, the new product is more useful and less expensive than the product displaced, and the barriers inhibiting trade are lower.

After consumption begins in other countries, the innovating country may still remain the sole producer of the new good, filling both domestic

2. Since new products are normally developed in response to domestic demand, consumption rarely occurs in imitating countries before it has in the innovating country. See Vernon, "International Investment," pp. 190–207; and Staffan Burenstam Linder, *An Essay on Trade and Transformation* (Wiley, 1961), pp. 87–91.

Figure 3-1. The International Diffusion Process[a]

(a) Innovating country

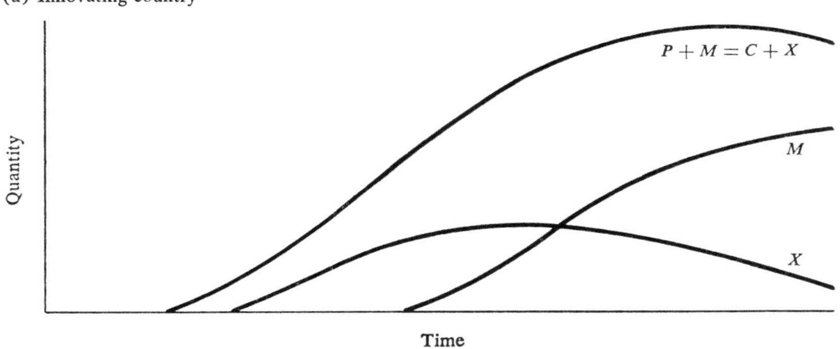

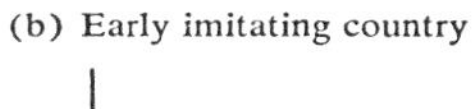

(b) Early imitating country

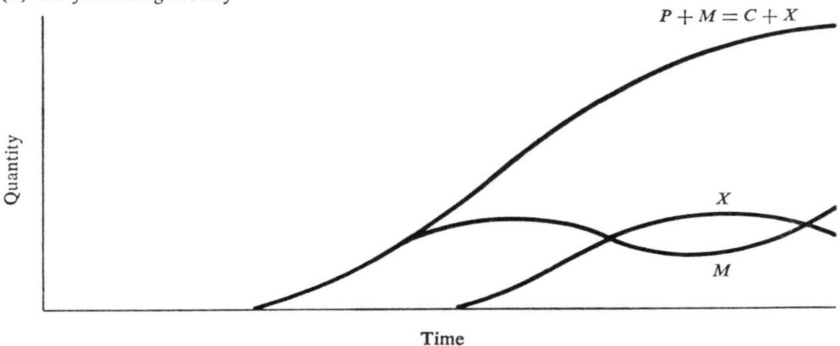

(c) Late imitating country

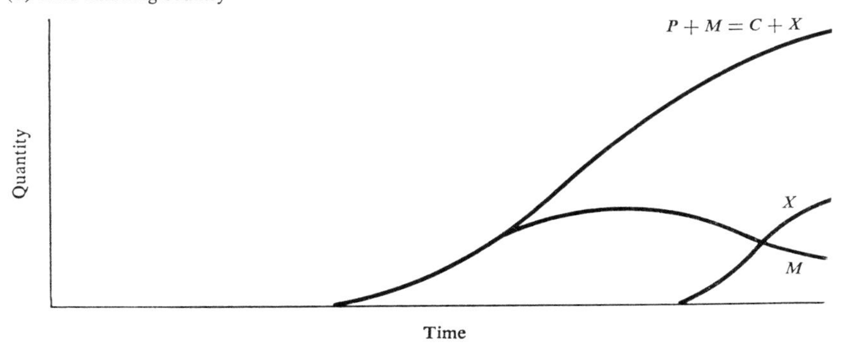

a. The X curve shows the level of a country's exports of a new product (or product produced with a new process); the M curve its imports. The $P + M$ curve is the sum of domestic production and imports. It indicates the total amount of the product available to the country for domestic consumption and exports $(C + X)$. The vertical distance between the $P + M$ curve and the M curve reflects the country's production and the vertical distance between the $C + X$ curve and the X curve its consumption.

and foreign demand, at least for a while. During this stage, the new technology is embodied in the product and transferred indirectly through exports. The transfer becomes complete when production begins abroad.

The interval between initial production in the innovating country and that in an imitating country is the "imitation lag." Like the demand lag, it varies with innovations and countries. Generally, it is shorter: the less technically sophisticated the new product, the less the need for follow-on research and development (R&D), the smaller the economies derived from scale and from producer learning, the greater the potential comparative advantage of the imitating country, the lower the barriers to its exports, the larger its domestic demand for the product, and the harder it is for imports to meet this demand.

The output of an early imitating country may replace imports, compete with the innovating country for markets in third countries, and even penetrate the domestic market of the innovating country. These trends, as Figure 3-1 illustrates, may reverse the initially unfavorable trade balance of the early imitating country in the new product and the initially favorable balance of the innovating country. Later, as other countries acquire the technology to produce the new good and gain a comparative advantage in its production, the early imitating country may suffer the same fate as the innovating country.

Figure 3-1 draws attention to several important features of the international diffusion process.

First, diffusion occurs on two levels—among users or consumers (demand) and among producers (supply). Within a country, diffusion on either level can proceed independently of the other as long as imports can satisfy demand and exports can absorb supply. For example, diffusion may never occur among producers if, by the time the country acquires the production technology, imports are cheaper than domestic products.

Public officials and others interested in the international diffusion of technology generally focus on diffusion in supply, emphasizing the effect of diffusion among producers on an imitating country's economy and balance of trade. Yet in some instances—polio vaccine is an example—diffusion among users can be more important.

Second, the international diffusion of new technology involves two steps: intercountry transfer of the technology (either directly or embodied in imports) and intracountry diffusion. Four aspects of diffusion can thus be examined in assessing country performance in the acquisition of new technology. The first is the speed with which a country initially tries a new product, or the demand lag. The second is how quickly the use of the product spreads among consumers after introduction into the domestic market, as indicated by the growth in the country's consumption. The third is the speed with which the country acquires the production technology from abroad, or the imitation lag. The fourth is how quickly domestic producers adopt the technology once it is successfully transplanted into the country from abroad, as indicated by the growth in the country's output.

Third, international trade and the diffusion of technology are closely intertwined and interacting. Trade can cause diffusion at the user level to occur sooner and more rapidly than otherwise possible. At the producer level, trade stimulates diffusion in countries with a comparative advantage in the new good and impedes diffusion in other countries.

In turn, the diffusion of technology affects the level and direction of trade. Comparative advantage shifts from one country to another when the technology needed to produce the new good is acquired by countries with lower factor costs. Importing nations may become exporters, and exporting nations importers. Within a country, diffusion on the producer level tends to reduce imports and increase exports, while on the consumer level it has the opposite effect.

Finally, when trade is blocked, diffusion of demand and production proceed at an equal pace. The determining factor of the rate of diffusion on the supply side is the country's ability to initiate and expand production of the new product, which depends upon the quality and quantity of domestic R&D resources, the accessibility of licensing and technical assistance agreements with foreign producers, the supply of factors of production, and so on. On the demand side, the important factors include the size of the domestic market, consumer tastes, and the propensity of potential users to become aware of and informed about new products. Without trade, unfavorable conditions in any supply or demand factor inhibit diffusion.

With trade, diffusion among users is still constrained by domestic demand conditions, but not by domestic supply conditions since imports are available as an alternative to domestic production. Similarly, diffusion among producers remains constrained by domestic supply factors, but is freed from domestic demand since exports are possible.

This raises an important consideration that must be taken into account when assessing company performance in diffusion on the producer level.

The relevant question for market structure policy is, Did the firms in a country adopt the new technology as soon as and to the extent justified by demand conditions? If trade is completely free, diffusion among producers is constrained in all countries by the same demand conditions—the size of the world market. But this is not the case if trade is blocked. Domestic demand for a new product may arise later and grow more slowly in some countries for many reasons, including consumer preferences or low per capita incomes. Consequently, firms may be slower to introduce and use the new technology for valid reasons. They should not be judged inept or deficient unless they are unresponsive to a favorable domestic market potential.

Diffusion of Semiconductor Technology

The performance of the United States, Britain, France, Germany, and Japan in adopting semiconductor technology is evaluated in this section. The emphasis is on diffusion in production, largely because this is the primary concern of these countries. Britain, for example, takes little comfort in the knowledge that most, perhaps even all, of the British demand for integrated circuits can be satisfied by imports from the United States. A second, equally compelling reason for focusing on diffusion at the producer level is that the data on production are more reliable and complete.

Intercountry Diffusion

The imitation lag reflects the rate of intercountry diffusion in production, and the demand lag the rate of intercountry diffusion in demand.

THE IMITATION LAG. Table 3-1 lists the date when each of the five countries began commercial production of the major semiconductor devices (identified in Table 2-1) along with the firm or firms responsible.[3] For lack of data, first laboratory models of the tunnel and Gunn diodes are indicated instead of first commercial production. Commercial production of the Gunn diode has not yet begun in several countries.

3. Also of interest are the imitation lags for the major process innovations of Table 2-2, but the data are not available. Indeed, just collecting the data for Table 3-1 presented many difficulties. Often it was difficult to determine exactly when experimental production ceased and commercial production began. Thus, conflicting claims concerning both the date and the firm responsible were encountered for some of the innovations.

Table 3-1. First Commercial Production of Major Semiconductor Devices and Firms Responsible, by Country, 1951–68[a]

Item	*Year*	*Firm*
Point contact transistor		
United States	1951	Western Electric
Great Britain	1953	Standard Telephones and Cables
		General Electric Company Ltd.[b]
		Associated Electrical Industries
France	1952	Société Européenne des Semiconducteurs
Germany	1953	Intermetall
		Siemens
		Valvo
Japan	1953	Sony
Grown junction transistor		
United States	1951	Western Electric
Great Britain	[c]	
France	1953	Société Européenne des Semiconducteurs
Germany	1953	Allgemeine Elektrizitäts Gesellschaft-Telefunken
		Siemens
		Valvo
Japan	1955	Nippon Electric
		Sony
Alloy junction transistor		
United States	1952	General Electric
		RCA
Great Britain	1953	Associated Electrical Industries
		General Electric Company Ltd.
France	1954	Radiotechnique Compelec
Germany	1954	Intermetall
		Valvo
Japan	1954	Fujitsu
		Sony
Surface barrier transistor		
United States	1954	Philco
Great Britain	1958	Semiconductors Ltd.[d]
France	[c]	
Germany	[c]	
Japan	1962	Fujitsu
Silicon junction transistor		
United States	1954	Texas Instruments
Great Britain	1958	Texas Instruments Ltd.
France	1960	Société Européenne des Semiconducteurs
Germany	1955	Valvo
Japan	1959	Nippon Electric
Diffused transistor		
United States	1956	Western Electric
		Texas Instruments
Great Britain	1959	Mullard
France	1959	Radiotechnique Compelec

Table 3-1 (*continued*)

Item	*Year*	*Firm*
Germany	1959	Siemens
		Valvo
Japan	1958	Nippon Electric
Silicon controlled rectifier		
United States	1956	General Electric
Great Britain	1957	Associated Electrical Industries
France	1960	Société Européenne des Semiconducteurs
Germany	1960	Allgemeine Elektrizitäts Gesellschaft-Telefunken
Japan	1960	Toshiba
Tunnel diode[e]		
United States	1958	RCA
Great Britain	1960	Standard Telephones and Cables
France	1960	Compagnie Générale des Semiconducteurs
Germany	1960	Allgemeine Elektrizitäts Gesellschaft-Telefunken
		Siemens
Japan	1957	Sony
Planar transistor		
United States	1960	Fairchild
Great Britain	1961	Ferranti
France	1963	Société Européenne des Semiconducteurs
Germany	1962	Intermetall
Japan	1961	Nippon Electric
		Fujitsu
Epitaxial transistor		
United States	1960	Western Electric
Great Britain	1962	Texas Instruments Ltd.
		Standard Telephones and Cables
France	1963	Société Européenne des Semiconducteurs
		Texas Instruments–France
Germany	1962	Intermetall
Japan	1961	Toshiba
		Sony
Integrated circuit		
United States	1961	Texas Instruments
		Fairchild
Great Britain	1962	Texas Instruments Ltd.
France	1964	Société Européenne des Semiconducteurs
		Compagnie Générale des Semiconducteurs
Germany	1965	Allgemeine Elektrizitäts Gesellschaft-Telefunken
		Siemens
Japan	1962	Nippon Electric
MOS transistor		
United States	1962	Fairchild
Great Britain	1964	Ferranti
France	1964	Compagnie Générale des Semiconducteurs
Germany	1965	Valvo
Japan	1963	Nippon Electric

Table 3-1 (*continued*)

Item	*Year*	*Firm*
Gunn diode[f]		
United States	1963	International Business Machines
Great Britain	1965	Standard Telephones and Cables
		Plessey
		Mullard
France	1965	Radiotechnique Compelec
Germany	1967	Valvo
Japan	1965	Nippon Electric

Sources: Christopher Freeman, "Research and Development in Electronic Capital Goods," *National Institute Economic Review*, No. 34 (November 1965), p. 64; Anthony M. Golding, "The Semiconductor Industry in Britain and the United States: A Case Study in Innovation, Growth and the Diffusion of Technology" (Ph.D. dissertation, University of Sussex, forthcoming); interviews with businessmen and government officials; correspondence with firms and trade associations; and various articles in the trade press.

a. A number of the firms have altered or completely changed their names over the last twenty years. To avoid confusion, the same company name (generally the most recent) is used for those firms that initiated the production of two or more devices and in the interim changed their names. Former names are given in Tables 4-1, 5-1, and 6-1 for American, European, and Japanese firms, respectively.

b. Not associated with the American firm.

c. Never produced in the country, or only in very limited quantities.

d. A joint venture of Plessey, with 51 percent of the equity, and the American firm Philco, with 49 percent of the equity.

e. The firms and dates indicated for the tunnel diode are for the first laboratory models. The first commercial production was by RCA and General Electric in the United States about 1960.

f. The firms and dates indicated for the Gunn diode are for the first laboratory models. First (successful) commercial production began in 1965 in Britain. The firm responsible was Associated Semiconductor Manufacturers, which at that time was two-thirds owned by Mullard and one-third by General Electric Company Ltd.

The figures below, based on the data in Table 3-1, indicate that the technology needed to produce new semiconductors has diffused swiftly to all five countries:

Average imitation lag, in years	*United States*	*Great Britain*	*France*	*Germany*	*Japan*
For all innovations	0.1	2.2	2.8	2.7	2.5
For all transistor innovations except surface barrier	0.1	2.0	2.8	2.7	2.1
For first 8 innovations in Table 3-1	0.1	2.6	3.0	2.4	3.4
For last 5 innovations in Table 3-1	0.0	1.6	2.6	3.0	1.2

Nearly 60 percent of the imitation lags are two years or less; more than 90 percent are four years or less.[4]

4. These figures exclude the imitation lags of the innovating countries, which by definition are zero, as well as the three cases in which a device was never produced in a country.

Britain, with an average lag of 2.2 years, appears to be the fastest imitator over the entire period. Japan is next with a lag of 2.5 years, followed closely by Germany and France. If the surface barrier transistor, which was never produced in France or Germany and which may not belong among the major semiconductor innovations in any event, is excluded, then Japan and Britain both have an average lag of about two years.[5]

A comparison of the average imitation lags for the first eight innovations, introduced in the 1950s, and for the last five innovations, introduced in the 1960s, indicates that Germany was the only country that did not reduce its average imitation lag in the 1960s. The French lag declined by five months, the British lag by a year, and the Japanese lag by more than two years, making that country the fastest imitator in the sixties.[6]

5. For comparison, the imitation lags for fourteen postwar innovations in synthetic rubbers and man-made fibers average 1.1 years for the United States, 5.4 years for Britain, 6.0 years for France, 4.4 years for Germany, and 6.3 years for Japan. For consistency, cases (which were much more numerous with synthetic materials) where a country never produced the product were excluded from the calculations. For the original data, see Hufbauer, *Synthetic Materials,* pp. 131-32.

The average imitation lags for the oxygen steel process, continuous casting of steel, special processes in papermaking, numerically controlled machine tools, shuttleless looms, and the float glass process are 1.4 years for the United States, 4.0 years for Britain, 4.5 years for France, and 4.0 years for Germany. The U.S. figure excludes continuous casting. If 1962 is the appropriate date for the first commercial use of this innovation in the United States, as suggested by Adams and Dirlam, the average U.S. lag is 2.8 years. See G. F. Ray, "The Diffusion of New Technology: A Study of Ten New Processes in Nine Industries," *National Institute Economic Review,* No. 48 (May 1969), pp. 40-83; and Walter Adams and Joel B. Dirlam, "Steel Imports and Vertical Oligopoly Power: Reply," *American Economic Review,* Vol. 56 (March 1966), p. 164.

6. The statistical significance of the British, French, German, and Japanese imitation lags and their intertemporal changes was assessed with regression analysis. Imitation lags were assumed to be related to the identity of the imitating country and the time the innovation was first introduced in the following manner:

$$Y_i = \sum_{j=1}^{4} a_{1j}X_{ji} + \sum_{j=1}^{4} a_{2j}X_{ji}T_i$$

where Y_i is the length in years of the ith imitation lag; X_{1i} is a dummy variable that equals one if Britain is the imitating country for the ith imitation lag and zero otherwise; X_{2i}, X_{3i}, and X_{4i} are similar dummy variables for France, Germany, and Japan, respectively; and T_i is a dummy variable valued at zero (one) for the eight innovations introduced in the 1950s and one (zero) for the five innovations introduced in the 1960s in the first (second) regression. When all in-

THE DEMAND LAG. Information on when new semiconductors were first used commercially in various countries is not available, so the average demand lags cannot be calculated in the same manner as were the average imitation lags. Nevertheless, some qualitative evidence about demand lags does exist.

Since none of the countries under consideration first produced semiconductors solely for export, demand lags cannot be longer than imitation lags. Moreover, the prevailing opinion in the industry is that increasing worldwide competition in semiconductors substantially shortened the demand lag during the 1960s. Among American firms the time between the introduction of new devices in the United States and their introduction in Europe apparently has greatly diminished.[7]

Intracountry Diffusion

How rapidly new semiconductor technology spreads among producers in a particular country is reflected by the ratio of semiconductor production to active component production. The faster this ratio moves from

---

dependent variables are dummy variables as in this equation, regression analysis is the same as analysis of variance.

The coefficients and standard errors (in parentheses below the coefficients) are shown below:

Regression number	a_{11}	a_{12}	a_{13}	a_{14}	a_{21}	a_{22}	a_{23}	a_{24}	R^2	*Number of observations*
1	2.57	3.00	2.43	3.37	−0.97	−0.40	0.57	−2.18	0.21	49
	(0.53)	(0.53)	(0.53)	(0.50)	(0.82)	(0.82)	(0.82)	(0.80)		
2	1.60	2.60	3.00	1.20	0.97	0.40	−0.57	2.18	0.21	49
	(0.63)	(0.63)	(0.63)	(0.63)	(0.82)	(0.82)	(0.82)	(0.80)		

They indicate (a) that the imitation lags for these four countries were significantly (at the 95 percent probability level) greater than zero for innovations introduced in both the 1950s and 1960s; (b) that, except for Japan, imitation lags have not significantly decreased or increased in the 1960s; and (c) that differences between the imitation lags of these four countries were not significant with the exception of the difference between the German and Japanese lags for innovations introduced in the 1960s. When the analysis was expanded to include the imitation lags for the United States (which with one exception are all zero), differences between the lags of the United States and the other four countries were found to be significant in both the 1950s and 1960s except for the difference between the Japanese and American lags for innovations introduced in the 1960s. None of these findings changed when the imitation lags for the surface barrier transistor were excluded from the analysis.

7. Organisation for Economic Co-operation and Development, *Electronic Components: Gaps in Technology* (Paris: OECD, 1968), pp. 114-15.

zero toward an upper limit of one,[8] the faster diffusion in production tends to proceed. Similarly, the ratio of semiconductor usage to active component usage reflects intracountry diffusion of demand as semiconductors are substituted for tubes and as new electronic products feasible only with semiconductors are produced.[9]

SEMICONDUCTOR VERSUS TUBE PRODUCTION. Figure 3-2 shows in section (a) the (value) ratio of semiconductor production to active-component production for each of the five countries for the years when data are available. The penetration of semiconductor technology among active-component producers proceeded swiftly in all countries during the 1950s, except possibly in Britain where the data are incomplete. Country ratios reach a peak or plateau at between 20 and 40 percent in the early 1960s, when increased capacity and new techniques, such as the planar and epitaxial processes, led to rapid price reductions.[10] After a pause of two

8. The complete substitution of semiconductors for tubes is not feasible, so the ceiling is actually less than one. It has, though, been rising as new developments expand the capabilities of semiconductor devices and reduce their costs.

9. Although other factors could conceivably affect these two ratios and impair their usefulness as a measure of intracountry diffusion of semiconductor technology, only one extraneous factor appears important enough to be taken into account when country ratios are compared: the variation between countries in the ratios accounted for by cathode ray and certain special purpose tubes for which semiconductors are not superior substitutes. This complication can be avoided for countries whose data permit these tubes to be excluded from the two ratios.

One other possible drawback in using the ratios to measure intracountry diffusion is that they do not distinguish between various types of semiconductors. The technology required to produce or use different devices varies greatly, and a country might have a high ratio simply because it produces or uses large volumes of relatively unsophisticated semiconductors. The effect of this limitation is reduced by measuring production and consumption by value rather than quantity. In the final section of this chapter differences between countries in the composition of semiconductor production are considered and attributed primarily to differences in demand conditions.

10. The number of semiconductors produced, though, continued to grow briskly. Had the ratio of the number (rather than the value) of semiconductor devices to the number of active components been considered in Figure 3-2(a), the peaks and plateaus of the early sixties would disappear. Also, all ratios would increase more rapidly over the entire period because semiconductor prices fell faster than tube prices. The ratio for Japan, in particular, would shift upward since that country has emphasized inexpensive semiconductors for consumer products.

Had the number of functions performed by semiconductors and tubes been compared in Figure 3-2(a), the ratios would increase at a still faster rate in the sixties because integrated circuits perform a number of functions, not just one. The average integrated circuit by the end of the decade probably performed at least

to three years, the ratios resume their upward climb, reaching 30 to 50 percent by 1968.

Initially, semiconductors displaced only receiving tubes, and even today their penetration into the market for special purpose tubes, though under way, is small compared to their penetration of the receiving tube market. In the case of cathode ray tubes, semiconductors are not yet viable substitutes. Hence a comparison of semiconductor production with receiving tube output is of interest. For international comparisons of intracountry diffusion, this ratio is superior to the other since it is not affected by irrelevant differences between countries in cathode ray and special purpose tube production. But unfortunately, data on the value of receiving tube production are available for only three countries.

Figure 3-2(b) illustrates this ratio for these countries. It portrays a slightly different picture of intracountry diffusion than Figure 3-2(a) shows. First, the American rate exceeds the Japanese in every year except 1958, indicating that diffusion did not occur faster in Japan than the United States in the latter half of the 1960s. The opposite impression is incorrectly created by Figure 3-2(a), because the production of special purpose tubes compared to that of receiving tubes is much greater in the United States than in Japan.

Second, much smaller disparities exist between countries in Figure 3-2(b). In part, this is because Germany and Britain, the two biggest laggards, are omitted. Nevertheless, the French ratio follows more closely that of the United States, seldom differing by more than ten percentage points, and as the overall level of penetration has risen, this difference has become relatively less important. The same is also true for the smaller but more changeable gap between Japan and the United States. Once irrelevant differences between countries in cathode ray and special purpose tube production are eliminated, Figure 3-2(b) indicates that the intracountry diffusion of semiconductor technology has proceeded at about the same pace in the three countries for which data are available.[11]

---

ten functions. This measure would particularly increase the ratios for the United States and Britain, the two most important integrated circuit producers.

11. The intracountry diffusion of semiconductor technology appears to have proceeded more rapidly in the countries considered than has the diffusion of most other innovations for which data are available. Studies on the diffusion of other innovations are summarized in Edwin Mansfield, *The Economics of Technological Change* (Norton, 1968), pp. 114–19. Also see Ray, "Diffusion of New Technology."

In making comparisons between the rate of substitution of semiconductors for

Figure 3-2. Value of Semiconductor Production, Five Countries, 1952–68

(a) Value as a percentage of active component production

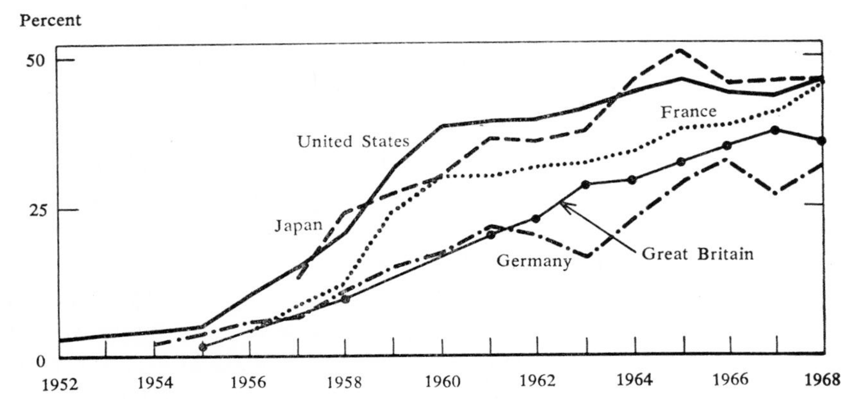

(b) Value as a percentage of receiving tube plus semiconductor production[a]

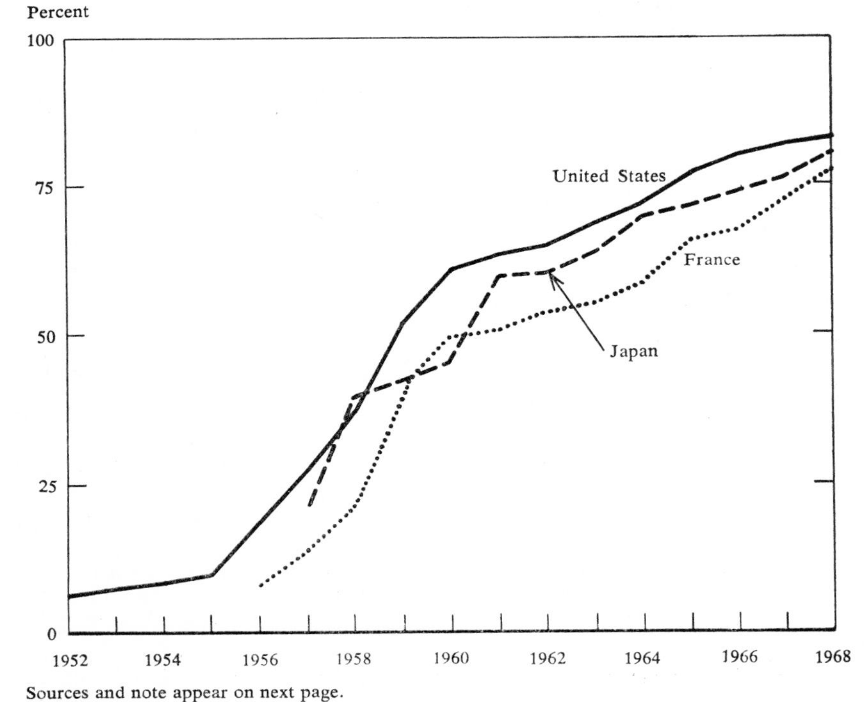

Sources and note appear on next page.

SEMICONDUCTOR VERSUS TUBE USAGE. How rapidly did producers of final electronic equipment substitute semiconductors for tubes and expand output of new products that semiconductors made feasible? Figure 3-3 indicates the penetration of semiconductors in total active-component

---

tubes and the rate of substitution of other innovations for the products or processes they replaced, it should be noted that differences do not entirely reflect differences in diffusion. Since semiconductors are not superior substitutes for electron tubes or even for receiving tubes in many applications, the upward movement of semiconductor production as a percentage of semiconductor plus tube production (or semiconductor plus receiving tube production) is retarded not only by the rate of diffusion but also by the fact that the equilibrium level is less than 100 percent. Indeed, when all electron tubes are considered rather than just receiving tubes, the equilibrium is considerably less than 100 percent.

To the extent that the equilibrium percentage is at a lower level and over time has moved upward more slowly for semiconductors than for the innovations with which semiconductors are compared, there is a tendency to underestimate semiconductor diffusion. Adjusting for this bias would provide additional support for the conclusion that the intracountry diffusion of semiconductor technology has been rapid compared to the diffusion of other innovations.

Figure 3-2, *notes*

Sources: *United States.* Data for tubes and discrete semiconductors: U.S. Department of Commerce, Business and Defense Services Administration, *Electronic Components: Production and Related Data, 1952–1959* (1960): *BDSA,* "Consolidated Tabulation: Shipments of Selected Electronic Components" (annual reports; processed) (title varied somewhat over the period). Monolithic integrated circuits (included in the 1962–68 production data) for 1963–68: *Electronic Industries Yearbook, 1969* (Washington: Electronic Industries Association, 1969), Table 55. The 1962 figure is an estimate. Before 1962 integrated circuit production was negligible.

Great Britain. Tube production for 1961–68 is estimated on the basis of VASCA (Electronic Valve and Semiconductor Manufacturers' Association) sales figures. Data for 1954 and 1958 are found in Board of Trade, *The Report on the Census of Production for 1958,* Part 59: *Radio and Other Electronic Apparatus* (London: Her Majesty's Stationery Office, 1961). The 1955 figure is an estimate based on 1954 tube production. Semiconductor data are from Anthony M. Golding, "The Semiconductor Industry in Britain and the United States: A Case Study in Innovation, Growth and the Diffusion of Technology" (Ph.D. dissertation, University of Sussex, forthcoming), and are estimates based on the statistics of VASCA indicating the sales of member companies, supplemented by information from firms and other organizations.

France. Production data for tubes and semiconductors are collected by the Fédération Nationale des Industries Electroniques (FNIE). Figures for 1968 are reported in *Rapport d'Activité, 1968* (Paris, FNIE); for 1964–67 in "L'Electronique en France: Statistiques et Réglementation" (Paris: FNIE, quarterly; processed); for 1959–63 in "Etudes et Documents: Statistiques Professionnelles de l'Electronique Française, 1959–1963" (Paris: FNIE, 1964; processed); and for 1956–58 in (U.S.) BDSA, *Electron Tubes and Semiconductors: Production, Consumption, Trade, Selected European Countries* (January 1960). The 1968 figures were adjusted to include taxes. Data before 1959 may not be completely consistent with later figures.

Germany. Data for tubes and semiconductors were provided by the Zentralverband der Elektrotechnischen Industrie, Frankfurt.

Japan. Production data for tubes and discrete semiconductors are from the Japanese Ministry of International Trade and Industry, as reported in BDSA, "Japanese Communications and Electronics Production" (semiannual report). Production data for 1966–68 also include monolithic integrated circuits, production of which was negligible in Japan before 1966. Data on integrated circuit production are from *Electronic Component Market Survey, Japan, 1969* (Tokyo: Science Newspaper Company and the Electric Apparatus Market Survey Committee), Table 10 (in Japanese; translated for author).

a. Separate data on receiving tubes are not available for Great Britain and Germany.

Figure 3-3. Value of Semiconductor Consumption, Five Countries, 1966

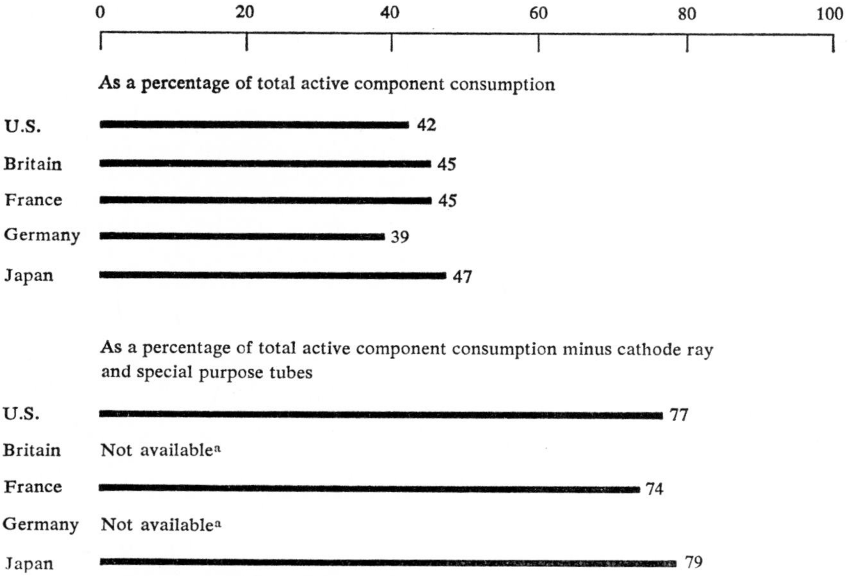

Sources: Semiconductor consumption data for Britain, France, and Germany for 1966 were provided by the Bureau d'Informations et de Prévisions Economiques, Paris, based on information obtained from government agencies and trade associations in these countries. Apparent consumption, determined by subtracting exports from production and then adding imports, was used to estimate semiconductor consumption in the United States and Japan and electronic tube consumption in all five countries. Trade and production data were obtained from the sources cited for Figures 3-2 and 3-5.

The limitations of apparent consumption data are noted below in Figure 3-5. Differences in the classification systems used for import, export, and production statistics, and other inadequacies in the data, at times required the use of liberal estimation procedures.

a. Data on the consumption of cathode ray and special purpose tubes in Britain and the consumption of special purpose tubes in Germany are not separately available.

consumption for all five countries in 1966. Ratios excluding cathode ray tubes and special tubes are also shown for those countries where the data are available. This removes the irrelevant effects of intercountry differences in the use of tubes that cannot be replaced with semiconductors. The results indicate that, at least in 1966, there were no large differences between countries in the intracountry diffusion of semiconductor technology on the user level.

Country Performance

The preceding measures of diffusion indicate that new semiconductor technology spread rapidly to and within the countries considered. Rarely do lags between the innovating and imitating countries exceed four years, and usually they are two years or less. Still in such a research-intensive industry where technology is advancing rapidly, a lag of only two years may place a country at a considerable competitive disadvantage. It is therefore important to determine whether diffusion is constrained primarily by supply or by demand conditions; in other words, whether firms use new semiconductor technology as soon as market opportunities arise. If they do not, because of inability to perceive the opportunities or to acquire the new technology, public policies that facilitate the entry of new firms or alter other market-structure characteristics might further stimulate the diffusion of semiconductor technology.

Although the impact of the many factors affecting diffusion cannot be precisely separated, several considerations suggest that diffusion has proceeded as quickly as justified by demand conditions in the countries considered and that differences between them arise primarily because their producers face different opportunities for using new semiconductor technology. The reasoning leading to this conclusion rests on a number of characteristics of semiconductor technology and markets that are only noted here since they are more fully described and supported in later chapters.

In considering diffusion among producers, the differences among countries in the size and composition of their semiconductor markets are important. The United States has by far the largest market (see Figure 3-5 below), and because of the great demand for military electronic equipment, government devices comprise a relatively substantial share of its market.[12] In contrast, the industrial market is most important in Europe and the consumer market in Japan, though the Japanese industrial market has become increasingly important in recent years.

Typically, a new semiconductor device is expensive, and consequently it is first used in military and other government equipment where performance takes priority over costs. As firms acquire production experience, costs fall. Several years after a device is first used in military equipment, its price is often low enough to permit significant penetration of the industrial market.[13] After several more years, the device is often cheap

12. For more on the composition of the American, European, and Japanese semiconductor and electronic markets, see pp. 89–92, 123–27, and 152–53.

13. The finding that military production improves competitiveness in the commercial market runs counter to much of the conventional wisdom on this subject. Semiconductor production, however, benefits greatly from learning economies, which are for the most part transferable from military to commercial production because of the close similarity between military and commercial semiconductor devices and production technologies. Learning economies and the shifting market pattern they generate are described more fully below. See pp. 85–87 and 89–92.

enough to compete in the consumer market where the performance-cost trade-off is lowest.

This shifting market pattern, coupled with the differences in the size and composition of various countries' semiconductor markets, causes the demand for most new semiconductor products in Europe and Japan to lag several years behind demand in the United States. With free trade, such differences in domestic demand need not delay the diffusion of semiconductor technology among producers in Europe and Japan, which could initially produce for the American market. But trade barriers do exist, impeding imports into the American market. These barriers are most restrictive early in the product cycle when semiconductors are used primarily in government equipment. Preferential treatment is extended to domestic producers on government purchases,[14] and close liaison between users and producers is most beneficial at this stage.

With time, semiconductor devices become more standardized and their properties better known, reducing the need for liaison. Also, discrimination in government procurement becomes irrelevant and price more important in the purchase decision as semiconductor devices move from the government to the industrial and consumer stages. Thus, the trade barriers isolating the American market from foreign producers, although not disappearing completely,[15] fall during the product cycle. By the consumer stage, substantial quantities of devices are imported into the American market, indicating that by this time diffusion among producers

14. The "Buy American" policy generally requires that foreign firms bid 6 percent (including transportation costs and duty) under the lowest bid by an American firm and 12 percent under the lowest bid by an American firm located in a labor-surplus area or qualifying as a small business concern. For balance-of-payments reasons, military procurement procedures have since 1962 adjusted foreign bids (including transportation costs but not duty) by upping them 50 percent. This adjustment procedure or the 6–12 percent rule, whichever is most beneficial for American companies, is then applied in evaluating bids. Furthermore, many military contracts are let on a sole-supplier basis and foreign firms may not have the chance to compete at all. See Organisation for Economic Co-operation and Development, *Government Purchasing In Europe, North America and Japan* (Paris: OECD, 1966); and Robert Skole, "Government Electronics: Federal Outlets Tough for Foreigners," *Electronics,* Vol. 41 (Dec. 9, 1968), pp. 119–24.

15. All semiconductor imports are subject to an ad valorem duty which in 1968, for example, was 11 percent. Also, other things being equal, equipment manufacturers generally prefer to buy from domestic semiconductor firms. This makes their supply lines easier to oversee. Finally, in some instances, foreign firms are prohibited from exporting to the American market by licensing agreements with American firms.

is no longer necessarily constrained by the domestic demand conditions of an imitating country.

While demand conditions account for a lag in diffusion on the producer level of two to four years between the United States and the other countries, supply conditions appear in most instances to necessitate a shorter lag. Semiconductor production requires no scarce minerals or other rare factors of production. Rather, comparative advantage is primarily determined by technology and wage rates. Technology can arbitrarily be separated into two components: technical ability to produce a particular device, a prerequisite for production, and all other know-how affecting factor productivity. In this industry, the latter is dominated by the substantial learning economies that accrue with production experience. Since these economies depend on demand, they should not be considered as a constraint on diffusion imposed from the supply side.

This means that the supply constraint is principally determined by a country's capacity to acquire the technical ability needed to produce new semiconductor devices. The countries considered are all advanced technologically. All have modern research facilities and well-trained scientists and engineers in the semiconductor field. Moreover, the nature of the technology makes industrial secrecy difficult to maintain, and the major innovating firms follow liberal licensing policies.[16] Competent semiconductor firms can normally duplicate new devices within six to twelve months. Thus, supply considerations, aside from demand-dependent learning economies, rarely delay diffusion in Europe and Japan at the producer level by more than a year; the relevant constraint in these countries generally comes from the demand side.

Supply factors cannot delay diffusion among semiconductor users any longer than among producers in a country. Indeed, electronic equipment manufacturers may have access to new devices through imports before domestic production starts. Supply considerations have delayed diffusion on the user level also by one year at most.

The opportunities for firms to use new semiconductor devices depend on the demand for final electronic equipment. Given that new semiconductors are usually used for several years in government equipment before they are used in commercial equipment, that the American market for government electronic equipment is large compared to the European and Japanese markets, and that American imports of government equip-

16. See pp. 73–77 and 118–20.

ment are restricted by the preferential treatment given domestic firms and other trade barriers, the opportunities for European and Japanese firms to use new semiconductors generally lag several years behind the opportunities of American firms. So demand, rather than supply, apparently is also the relevant constraint on diffusion of semiconductor use.

The fact that the countries considered have used new semiconductor technology and devices as soon as demand produced the necessary opportunities does not mean that the established firms have always been adept at realizing these opportunities. But it does imply that when established firms have failed to introduce and use new technology as quickly as warranted by demand, the market structure of the semiconductor industry in these countries has allowed new companies to enter the industry and capture an important share of the market. Subsequent chapters will show that the type of firm responsible for rapid diffusion on the producer level in the United States, Europe, and Japan has varied greatly.

Semiconductor Trade and Production

This section first develops a simple model of semiconductor trade, which is then used to help analyze various aspects of semiconductor trade and production. For reasons given in the previous section, the model assumes, first, that demand conditions determine the imitation lags of the countries considered, and second, that once the ability to produce a semiconductor device is acquired, a country's comparative costs depend on learning economies and wage rates. Other factors are presumed to have an insignificant effect on costs.

The model also assumes that barriers to trade exist and that they vary in their restrictiveness over the life of a semiconductor device. As noted, the barriers protecting the American market fall in intensity over the product cycle, and by the consumer stage American imports are sizable. The European and Japanese markets are also protected by trade barriers,[17] but their impact varies in a different manner over the product cycle. Imports of new devices used largely in government products are restricted less than those for industrial and consumer goods since many of

17. These barriers differ in certain respects from the American barriers. For example, the preferential treatment given domestic producers on government purchases is often informal, rather than statutory as it is in the United States. Tariffs on semiconductors are somewhat higher than the American tariff.

the new devices are not yet produced in these countries or are produced only in limited quantities. Overall, the model assumes that the fraction of world production of a semiconductor device entering international trade tends to fall when the device penetrates the industrial market and when other countries besides the United States undertake significant production, and that this trend is reversed in the consumer stage primarily because of increased American imports. This hypothesized evolution in the propensity to trade over the product cycle is illustrated in the bottom half of Figure 3-4.

The process envisioned in the model by which comparative advantage in the production of a new semiconductor device shifts from one country to another is portrayed in the upper section of Figure 3-4. The United States, because of its large government market and the government-industrial-consumer shifting market pattern followed by most semiconductor devices, is expected to be the first producer of a new device. Initially, production costs are high, but they fall over time as American firms acquire production experience.[18]

Declining costs at some point permit the penetration of the industrial market, and at this time significant demand for the device arises in Europe, where the industrial electronic market is most important. Firms there will undertake production as soon as demand reaches the level necessary to assure that the discounted stream of profits anticipated over time is positive. The demand needed to induce domestic production is usually higher in Europe (and other imitating countries) than the level originally needed in the United States, because European electronic equipment makers can import new semiconductors from American firms and even benefit from the lower prices made possible by the U.S. producers' experience.

Since some of the learning from past production is not transferable, costs initially are higher in European firms than in the experienced American firms, as Figure 3-4 illustrates. (Because some learning *is* transferable and because wage rates are lower in Europe, initial production costs there are lower than costs were at first in the United States.)

18. Figure 3-4 shows unit costs falling at a decreasing rate over the entire product cycle. Since learning economies in this industry depend at least in part on cumulative output (see pp. 85–87), costs may decline at an increasing rate early in the product cycle when production is rapidly expanding. Eventually, though, costs must decline at a decreasing rate because cumulative production cannot for long increase at a geometric rate.

Figure 3-4. International Evolution of Production Costs for Semiconductor Devices and the Propensity to Trade over the Product Cycle

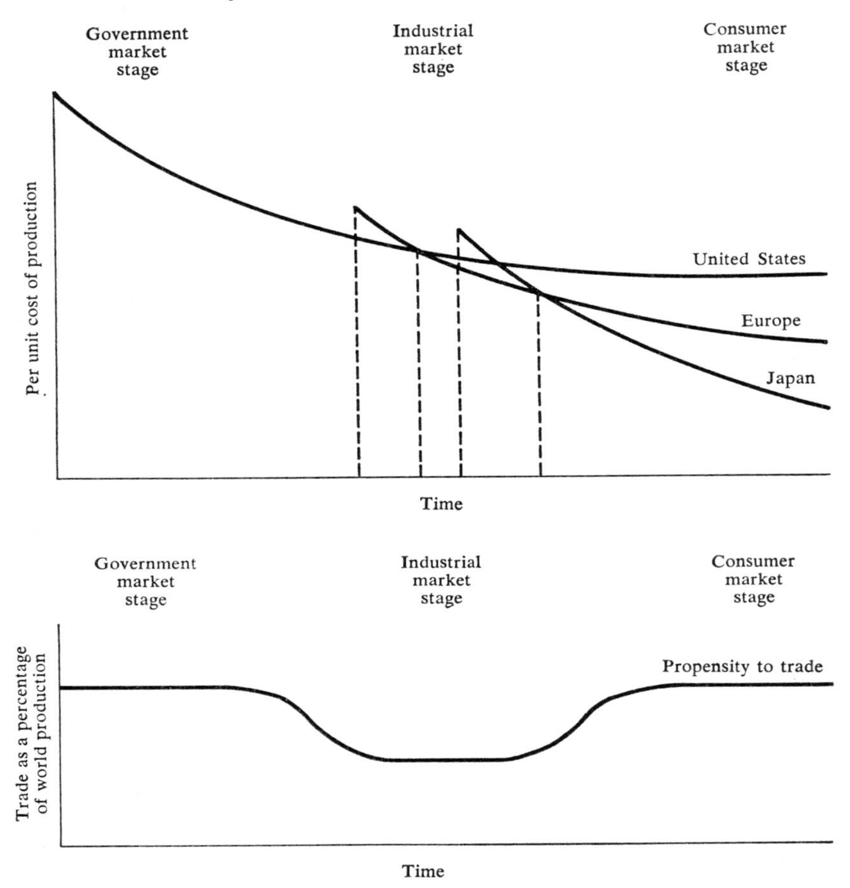

Despite the initial cost disadvantage, European firms will undertake production once sufficient demand arises in their home markets for two reasons.[19] First, in domestic markets trade barriers partly offset the lower costs of American firms. Second, the relative advantage that the latter derive from learning economies tends to diminish as European firms acquire production experience, and at some point this advantage is com-

19. Although the threshold level of demand needed to induce domestic production conceivably might never arise, this (as Table 3-1 suggests) has seldom been the case in the semiconductor industry.

pletely offset by the lower wage rates in Europe. Comparative advantage then shifts to the European firms (unless it has already shifted to firms in another country like Japan with even lower wage rates). Thus, although production is at first unprofitable, it generates learning economies, lower costs, and later profits that offset earlier losses.

The demand needed to support Japanese production arises at least by the time a device is used for consumer products. With the rapid growth of the Japanese market for industrial electronic equipment in the sixties, the production of many semiconductor devices now begins at about the same time as in Europe. In Figure 3-4, Japan is shown undertaking production shortly after Europe.

Initially, the Japanese production costs are likely to exceed the costs of the experienced foreign producers (though again, because of lower wage rates and the transferability of some learning, Japanese costs are lower than the initial costs of foreign producers). But as Japanese firms gain production experience, costs fall and eventually Japan acquires a comparative advantage. If Japanese firms begin producing a semiconductor device before European firms, as they have on occasion, it is very difficult for European firms ever to acquire a comparative advantage. To do so requires that, once the European firms begin, they produce on such a large scale that the learning economies generated offset not only the learning economies of the Japanese producers but also the advantage the latter derive from lower wage rates.

In the model developed here, comparative advantage in the production of a semiconductor device shifts from one country to another during the product cycle, but at any given time only one country is expected to be exporting the device. Because of trade barriers, other countries may be producing the device for the home market. When the model is applied to the semiconductor industry as a whole, which embraces many devices at various stages of the product cycle, it anticipates that the countries considered will both import and export semiconductors. Moreover, since these countries have the largest electronic equipment industries, they will trade primarily with each other. Trade data confirm these predictions.[20]

The model also explains the considerable differences among countries in the composition of semiconductor exports, imports, and production.[21]

20. See Figure 3-5 and the sources noted there.

21. Although semiconductor trade and production data do not identify individual devices, they are often sufficiently disaggregated to reflect the comparative importance of devices for government, industrial, and consumer products. For ex-

American exports and production are heavily weighted with relatively new and advanced devices used in military and industrial equipment, while American imports are mainly consumer devices. Conversely, Japanese production and exports are comprised largely of consumer devices, although this is changing as the industrial electronics market becomes more important there. Japanese imports are primarily technically advanced devices. European production and exports focus on industrial devices, though consumer devices are important too, since these countries have been less willing than the United States to import consumer devices from Japan. European imports are also primarily technically advanced devices.

The important factors determining a country's balance of trade in semiconductors and their interaction over time are identified by the model. Although learning economies postpone the day, once the technology diffuses abroad and countries with lower wage rates begin producing a semiconductor device, the United States soon loses its comparative advantage. In order to maintain a favorable balance of semiconductor trade, the country must continually introduce new and better devices. So far the United States has successfully done this. Figure 3-5, which shows semiconductor imports, exports, production, and consumption for the five countries over time,[22] indicates that the United States has enjoyed a favorable balance of semiconductor trade over the entire period for which data are available.[23] Moreover, this favorable balance has im-

ample, a low ratio of silicon transistors to germanium transistors or of integrated circuits to all semiconductors indicates that consumer devices are a relatively important part of a country's semiconductor exports, imports, or production. Sources with production and trade data for the countries considered are noted with Figures 3-2 and 3-5.

22. The postulated evolution in country production, consumption, exports, and imports portrayed in Figure 3-1 for a single semiconductor device deviates from the pattern shown in Figure 3-5 because the latter considers the entire semiconductor industry, which is composed of many semiconductor devices at various stages of the product cycle. Though not comparable, these two figures are related. For each semiconductor device there is a Figure 3-1; aggregating these figures for all devices at various points in time generates Figure 3-5.

23. The favorable trade balances noted here for the United States and Japan and the unfavorable balances noted for the European countries might conceivably be reversed if the semiconductors exported and imported in final electronic equipment were included in semiconductor trade data. This appears unlikely, though, given the strong export position of Japan in consumer electronic products and of the United States in military and industrial electronic equipment. Regarding the latter, see Christopher Freeman, "Research and Development in Electronic Capital Goods," *National Institute Economic Review,* No. 34 (November 1965), pp. 40–91.

proved in recent years, increasing from under $16 million in 1960 to over $130 million in 1968.

The export performance of an imitating country, according to the model, is a function of the interval between the time it begins production and the time another country with lower wage rates begins[24] and of the propensity to trade during this interval. Given that Japan generally undertakes the production of new devices soon after (or even before) the European countries, the model predicts (and Figure 3-5 confirms) that the European countries have unfavorable balances of semiconductor trade. The fact that trade is most restricted at the industrial stage of the product cycle, when the European countries are likely to be exporting, is an additional reason for their poor trade performance.

For Japan, a favorable balance of semiconductor trade is anticipated since the country enjoys a comparative advantage at a time when trade is relatively less restricted. Moreover, its imitation lag has shortened in the sixties, which tends to lengthen its export interval. Figure 3-5 indicates that Japan has had a favorable trade balance over most of the period considered, but it also shows that its balance has greatly declined in recent years. While Japan has been squeezing the export interval of the European countries, its own interval has been shortened by the initiation of semiconductor production in such countries as Hong Kong and Taiwan, whose wage rates are well below those of Japan.[25] Cost curves for these countries lie to the right and eventually below the curves shown for the advanced countries in Figure 3-4. Unlike the advanced countries, the imitation lags of these developing countries are determined by supply rather than demand conditions.

The model recognizes the presence of barriers to semiconductor trade and considers how their effectiveness varies over the product cycle, but it does not indicate how restrictive they are in general. Some evidence on this is provided by Figure 3-5, which shows that exports and imports are

24. An imitating country (as Figure 3-4 illustrates) will normally not possess a comparative advantage at the time it begins producing a new semiconductor device. But presumably, in order to capture foreign markets it is willing to sustain a greater loss on each device at the time it begins production than are its foreign competitors, because the production experience thereby acquired increases its cumulative output and lowers its costs by a greater percentage than is the case for its experienced foreign competitors. Thus, an imitating country should begin exporting a device soon after it starts production and before it has acquired a comparative advantage. Similarly, it should cease exporting soon after another country with lower wage rates undertakes production even though for a while it may still possess a comparative advantage.

25. The impact of these developments is considered further in Chapter 7.

Figure 3-5. Semiconductor Production, Consumption, Exports, and Imports for Five Countries, 1952–68[a]

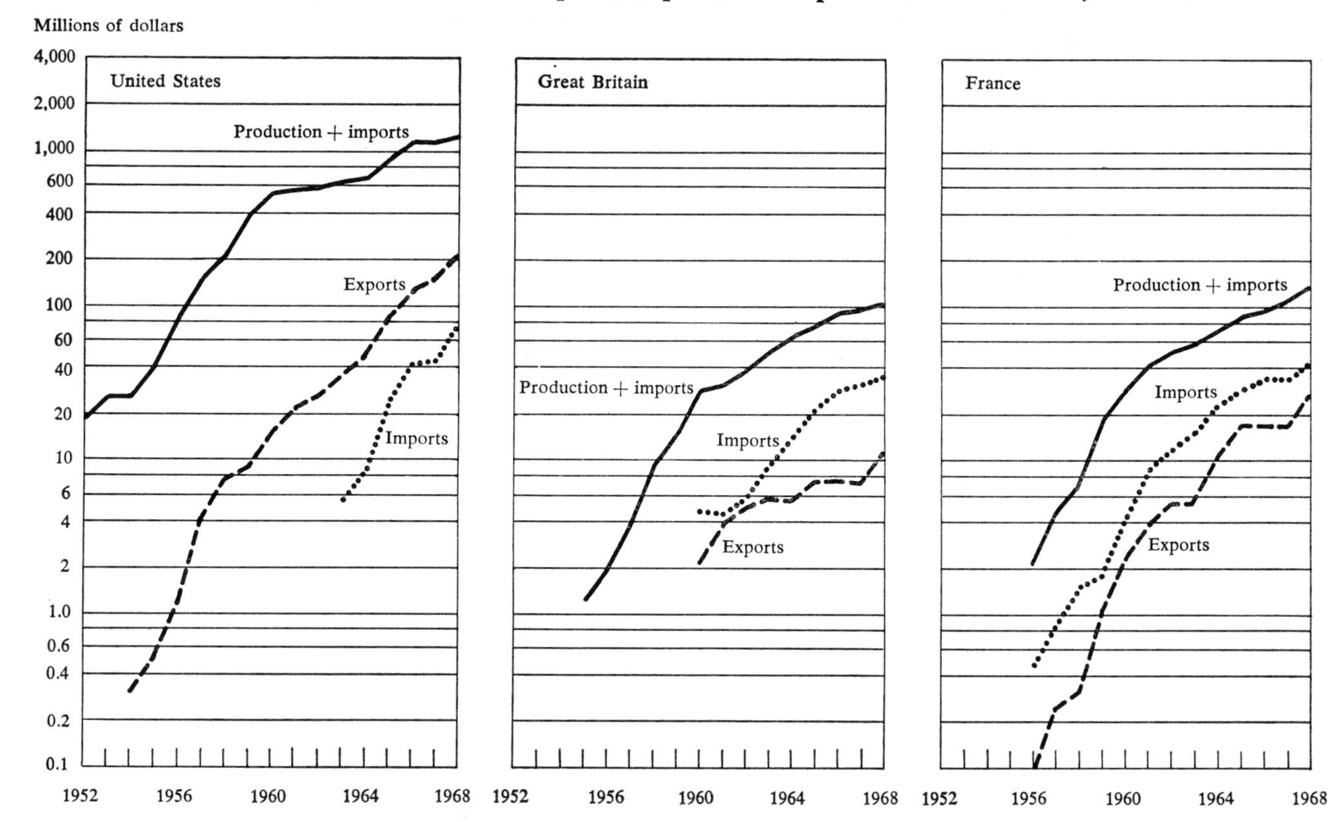

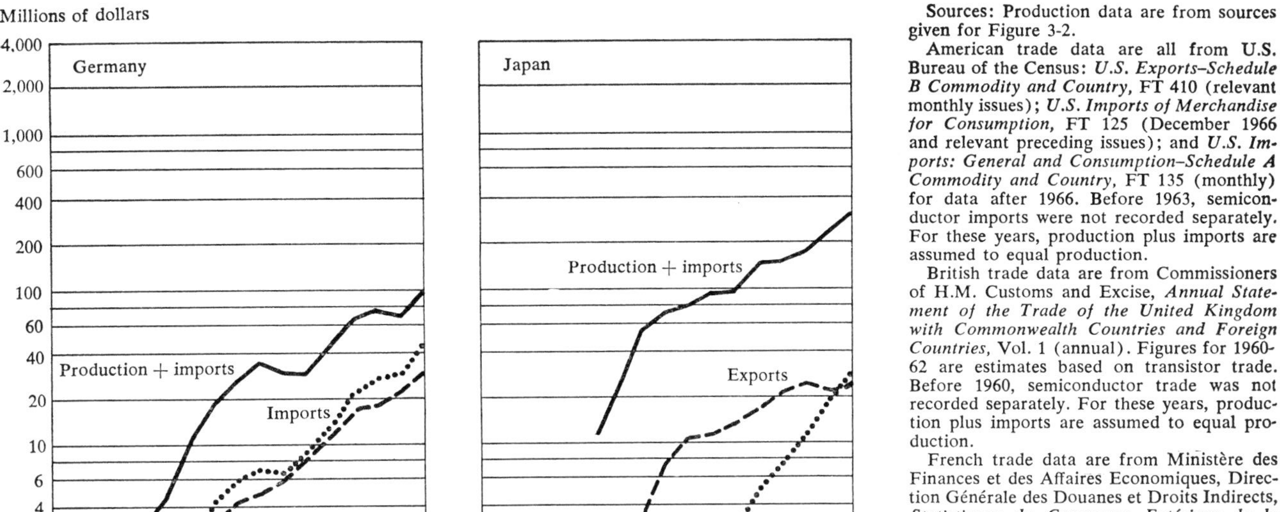

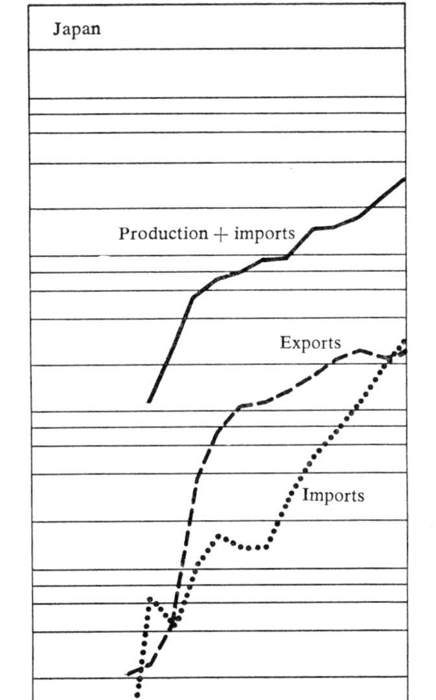

Sources: Production data are from sources given for Figure 3-2.

American trade data are all from U.S. Bureau of the Census: *U.S. Exports–Schedule B Commodity and Country,* FT 410 (relevant monthly issues); *U.S. Imports of Merchandise for Consumption,* FT 125 (December 1966 and relevant preceding issues); and *U.S. Imports: General and Consumption–Schedule A Commodity and Country,* FT 135 (monthly) for data after 1966. Before 1963, semiconductor imports were not recorded separately. For these years, production plus imports are assumed to equal production.

British trade data are from Commissioners of H.M. Customs and Excise, *Annual Statement of the Trade of the United Kingdom with Commonwealth Countries and Foreign Countries,* Vol. 1 (annual). Figures for 1960–62 are estimates based on transistor trade. Before 1960, semiconductor trade was not recorded separately. For these years, production plus imports are assumed to equal production.

French trade data are from Ministère des Finances et des Affaires Economiques, Direction Générale des Douanes et Droits Indirects, *Statistiques du Commerce Extérieur de la France* (annual).

German trade data were obtained from the German Federal Republic, Statistisches Bundesamt, Wiesbaden. Figures for 1966–68 are reported in *Aussenhandel, Reihe 2: Spezialhandel nach Waren und Ländern* (monthly). Multicrystal power rectifiers are excluded from the data. Before 1963, monocrystal power rectifiers are also excluded because they were reported with multicrystal power rectifiers.

Japanese trade data are from Ministry of Finance, *Japan Exports and Imports: Commodity by Country* (monthly). Title varies in earlier years.

a. The difference between the production-plus-imports curve and the imports curve indicates a country's production. Similarly, the difference between the production-plus-imports curve and the exports curve indicates its consumption since production plus imports equals consumption plus exports for each country. Consumption is apparent consumption derived from production, import, and export data. It is the least accurate of the four statistical series shown in the figure since it contains all errors found in the other three. Moreover, product classes for production, import, and export data are not commensurate in all cases.

a small fraction (rarely more than 10 percent) of production and consumption in both the United States and Japan. Even for the European countries, despite the relative freedom of intra-European trade, domestic production and consumption are larger than imports and exports.

The low volume of semiconductor trade has several implications. First, it suggests that trade is not an important channel for the diffusion of new semiconductor technology on the user or consumer level. But this conclusion must be qualified. Many semiconductor devices are exported in final electronic equipment and consequently are not recorded in semiconductor trade statistics. As Chapter 6 shows, most of Japan's transistors have been exported in transistor radios and other final goods. Similarly, American computers and military equipment shipped abroad contain many semiconductor devices. The ultimate users of such equipment benefit from advanced American semiconductor devices, though foreign electronic equipment producers do not. Even the diffusion of new semiconductor technology among the latter may be understated by the trade data. American exports, which are primarily technologically advanced devices, exceed the combined exports of the other four countries shown in Figure 3-5. This, coupled with the fact that an unknown portion of intra-European trade involves advanced devices, suggests that the ratio of trade to production of advanced devices may be higher than the ratio for all semiconductors.[26]

The low volume of semiconductor trade also indicates that diffusion on the producer level is primarily constrained by the size of a country's domestic market. The United States, with the largest electronic equipment market, is also the largest semiconductor producer. Japan, with a

26. Also, deficiencies in the data bias the trade figures downward. Some firms deliberately undervalue semiconductor shipments in order to reduce import duties. In addition, trade figures for some countries exclude semiconductor parts. This omission was probably not important during the fifties, but recently trade in diffused silicon slices has grown greatly. Preparing these slices constitutes an important and technically difficult part of integrated circuit production and the batch fabrication of discrete devices.

While understating trade, the data probably overstate production. In some countries, at least, double counting is known to occur. Companies occasionally sell unlabeled semiconductors to other semiconductor firms, which then resell them under their own trade names. Such transactions arise when a preferred supplier finds it cannot meet an order on time or in the agreed quantity and, rather than jeopardize future orders, buys devices from a competitor, perhaps even losing money in the process. The tied contract, whereby a firm agrees to supply as a part of a larger order some devices it does not produce, also contributes to the trade in unlabeled semiconductors.

sizable consumer market and a growing industrial market, is the second biggest semiconductor producer, though rarely does Japanese production exceed 20 percent of the American figure. Britain and France are the most important European producers,[27] but their output is less than one-tenth that of the United States. Thus in production as well as trade, the United States has been able to capitalize on semiconductor technology more than has Japan, and Japan in turn more than the European countries.

Conclusions

A new product or process is generally introduced in an advanced country where R&D resources are plentiful and demand conditions most favorable. In time, it disseminates. Usually the first imitators are other advanced countries, and less developed countries with lower wage rates later imitators.

The international diffusion of new technology proceeds on two levels: among the users of the good produced with the new technology and among the producers. At both levels, diffusion involves first the intercountry transfer of technology and then intracountry diffusion.

In the absence of international trade, diffusion among both users and producers is constrained by domestic supply and demand conditions. Trade frees diffusion of use from the domestic supply conditions and diffusion of production from the domestic demand conditions. Thus diffusion need not occur at the same pace on both levels.

An examination of imitation lags, the ratio of semiconductor to active-component production, demand lags, and the ratio of semiconductor to active-component consumption indicates that semiconductor technology diffused rapidly in the United States, Britain, France, Germany, and Japan. Rarely do the lags between countries exceed four years; in most instances they are two years or less.

The lags that do occur arise because companies face different domestic demand conditions and are kept out of foreign markets by trade barriers. On the supply side, difficulties in acquiring and using new semiconductor technology apparently do not retard diffusion. In other words, the diffu-

27. Since 1968, the last year for which data were available for this study, German semiconductor consumption and production have grown greatly. As a result, Germany may no longer be the smallest of the three European producers.

sion of semiconductor technology has proceeded as rapidly as warranted by demand conditions in the five advanced countries. This does not necessarily mean, as the following chapters will confirm, that the established firms in these countries always respond favorably to the opportunities created by new semiconductor technology. But it does imply that the market structure of the semiconductor industry in these countries is such that established firms are promptly disciplined or replaced when they fail to act quickly.

Despite the swift diffusion of semiconductor technology, the ability to use this technology to promote domestic production and to improve trade balances has varied greatly among the five countries. Because of trade barriers, the semiconductor production of a country is largely, though not completely, constrained by the size of its final electronic equipment market. As countries with lower and lower wage rates acquire the necessary technical ability and production experience, comparative advantage shifts to them. The interval during which a country can export depends not only on how quickly it acquires the necessary technology and undertakes production, but also on how quickly other countries with lower wage rates follow. The rapid diffusion of semiconductor technology to Japan curtails and in some instances even eliminates the period when the European countries enjoy a comparative advantage. As a result, these countries have unfavorable balances of semiconductor trade. The diffusion of semiconductor technology to Hong Kong and other countries with low wage rates has reversed Japan's favorable trade balance in recent years. On the other hand, the United States, despite the swift diffusion of semiconductor technology abroad, has managed to maintain and even enhance its trade balance by continually introducing new semiconductor innovations.

CHAPTER FOUR

The United States

THE DIFFUSION of semiconductor technology described in the previous chapter was not fortuitous but rather was produced by the conscious efforts of individual firms responding to economic forces. The next three chapters examine and contrast the types of firms and the economic factors responsible for the diffusion of this technology in the United States, Europe, and Japan. They also consider company contributions in advancing semiconductor technology, since in practice the development of new technology and its diffusion are closely intertwined.

This chapter focuses on the American scene.

Types of Firms

Semiconductor diodes and rectifiers were produced before World War II, but the industry received its major thrust forward after the war with the invention of the transistor. This development quickly captured the interest of the scientific community, if not of the general public,[1] and greatly increased the number of solid-state physicists and other specialists working in the semiconductor field. The result was a self-reinforcing spiral of advances in technology and growth in production.

Since the industry, though it existed and even made some progress earlier, was really launched by the advent of the transistor, we begin our investigation of semiconductor firms with 1951, the year the transistor was commercially introduced. We separate firms that have produced semiconductors since that time into three categories.

The first contains just one firm, American Telephone and Telegraph,

1. Bell Laboratories, in announcing the invention of the transistor in 1948, tried to avoid exaggerating its immediate significance since no one knew how long the development of a commercial device would take. As a consequence, the story covering one of the major inventions of the century ran on page 46 of the *New York Times*.

with Bell Laboratories, its research arm, and Western Electric, its manufacturing arm. For several reasons, AT&T merits its own category. First, Bell Laboratories has earned a special place in the annals of the semiconductor industry. It produced the original transistor and a disproportionately large share of the other major product and process innovations identified in Tables 2-1 and 2-2. These contributions reflect an enormous research and development (R&D) effort in the semiconductor field. Moreover, as later sections of this chapter indicate, the company has played a unique role in establishing the industry's liberal licensing policies and attitudes toward interfirm mobility of scientists and engineers, which have facilitated the entry of new firms. Finally, AT&T is the only firm enjoined from selling semiconductors in the commercial market. It surrendered this right by agreeing to the 1956 consent decree that culminated a lengthy antitrust case against the company.[2] It is free, however, to sell semiconductors in the military and space markets and to produce for its own needs.

The second category includes the receiving tube firms. In the early fifties, General Electric, RCA, and Sylvania were the three largest receiving tube producers, and together accounted for between 70 and 80 percent of American production. Philco, CBS (Columbia Broadcasting System), Raytheon, Tung-Sol, and Westinghouse were the only other manufacturers of any significance. Although all of them are large diversified electrical houses whose growth, profits, and survival do not depend on any one product, the rise of semiconductor technology has directly threatened their receiving tube operations.

The third category covers what we call the new firms. To some extent this is a misnomer. While all companies were new at transistor manufacture and the associated semiconductor technology in the early fifties, many in this category were not new in the sense of having recently been founded. However, Western Electric and the receiving tube firms were among the first to produce transistors commercially; by comparison nearly all the other firms that entered the industry were newcomers. Another essential distinction is that firms in this category were new to the active-component field or at least had no vested interests in receiving tubes to protect.

This third category, containing far more companies than the second, is

2. *United States* v. *Western Electric Company, Incorporated, and American Telephone and Telegraph Company,* Civil Action 17-49, *Final Judgment,* Jan. 24, 1956. In return for this and other concessions, AT&T persuaded the Justice Department to drop its efforts to sever Western Electric from the AT&T system.

comprised of three subgroups. The first includes firms like Hughes, Motorola, and IBM, which were large and diversified companies when they entered the semiconductor industry. Often these firms are producers of final electronic equipment or systems. A few manufacture semiconductors solely for in-house use, but most sell semiconductors outside the firm.

The second subgroup embraces small companies that were engaged in other branches of electronics or in other industries at the time they began making semiconductors. Texas Instruments, for example, when it first considered entering the semiconductor industry in 1949, was a small geophysical services company with annual sales of less than $6 million.

The last subgroup contains companies established to manufacture semiconductors. Most were founded by scientists and engineers who left Bell Laboratories and other established semiconductor firms. Transitron, for example, was formed in 1952 by two brothers, one a businessman and the other a solid-state physicist who had previously worked for Bell Laboratories.

Not all firms fall neatly into one of these three subclasses. An example is Shockley Laboratories, established by William Shockley, who shared a Nobel Prize with John Bardeen and Walter Brattain for inventing the transistor. Shockley left Bell Laboratories about 1954, recruited young scientists from industry and universities, and began searching for new semiconductors to develop and manufacture. Although sharing many features of firms in the last subclass, Shockley Laboratories obtained initial financial backing by becoming a wholly owned subsidiary of Beckman Instruments. A somewhat similar situation occurred a few years later when eight of Shockley's scientists decided to strike out on their own. Fairchild Camera and Instrument agreed to finance their efforts for eighteen months, and in return received an option, which it eventually exercised, to buy the new firm.

The years in which Western Electric, the eight receiving tube producers, and thirty-six of the new firms engaged in commercial transistor production are shown in Table 4-1. Except for Signetics, an important integrated-circuit manufacturer that does not produce transistors, the table includes all of the important semiconductor firms. By considering only transistor rather than all semiconductor producers, firms that produced semiconductor diodes and rectifiers in the early 1950s with methods predating the transistor and the new technology it precipitated are excluded.

Except for IBM, which produces transistors solely for in-house use, all

Table 4-1. Transistor Firms in the United States and Years They Were Active in the Semiconductor Industry, 1951–68[a]

Transistor firm	*1951*	*1952*	*1953*	*1954*	*1955*	*1956*	*1957*	*1958*	*1959*	*1960*	*1961*	*1962*	*1963*	*1964*	*1965*	*1966*	*1967*	*1968*
Western Electric[b]	o	o	o	o	o	o	o	o	o	o	o	x	x	x	o	o	o	o
General Electric	o	x	x	x	x	x	x	x	x	x	x	x	x	x	x	x	x	x
Raytheon	o	x	x	x	x	x	x	x	x	x	x	x	x	x	x	x	x	x
RCA	o	o	o	x	x	x	x	x	x	x	x	x	x	x	x	x	x	x
Bogue Electric[c]		x	x	x	x	x	x	x	x	x	x							
Clevite[d]		x	x	x	x	x	x	x	x	x	x	x	x	x	o			
Motorola[e]		o	o	o	o	x	x	x	x	x	x	x	x	x	x	x	x	x
National Union Electric		x	x	x	x	o	o											
Columbia Broadcasting System[f]			x	x	x	x	x	x	x	x	x							
Radio Receptor[g]			x	x	x													
Sylvania			x	o	x	x	x	x	x	x	x	x	x	x	x	x	x	o
Texas Instruments			x	x	x	x	x	x	x	x	x	x	x	x	x	x	x	x
Transitron			x	x	x	x	x	o	x	x	x	x	x	x	x	x	o	x
Tung-Sol[h]			o	o	x	x	x	x	x	x	x	x	x					
Westinghouse			x	x	o	x	x	x	x	o	x	x	x	x	o	x	x	x
Amperex[i]				o	x	x	x	x	x	x	x	x	x	x	x	x	x	x
Honeywell[j]				x	x	x	x	x	x	x	x	x	x	x	o			
Philco-Ford[k]				x	x	x	x	x	x	x	x	x	x	x	o	x	x	x
Thompson Ramo Wooldridge[l]				o	x	x	o	o	x	x	x	x	x	x	x	x	x	x
General Instrument[m]					x	x	x	x	x	x	x	o	x	x	x	x	x	x
Hughes					x	x	x	x	x	x	x	x	x	x	x	x	o	x
Sprague Electric					x	x	x	x	o	x	x	x	x	x	x	x	x	x
Delco Radio[n]						o	x	x	x	x	x	x	x	x	x	x	x	x
Nucleonic Products						x	x	x	o	x	x	x	o	x	x	x	x	o
Bendix							x	x	x	x	x	x	x	x	x	x	x	x
Hoffman							x	o	x	x	x	o	o	o	o	o	o	o
Fairchild								o	x	x	x	x	x	x	x	x	x	x
Industro Transistor								x	x	x	x	x						
Sperry Rand[o]								x	x	x	x	x	x	x	x	x	x	
Rheem[p]									x	x	x							
National Semiconductor[q]									o	o	x	x	x	x	x	x	x	x
Unitrode[r]									x	x	o	x	x	x	x	x	o	x
U.S. Transistor									o	x	x	x						
Crystalonics[s]									o	x	x	x	x	x	x			
Silicon Transistor[t]										x	x	x	x	x	x	o	x	x
Micro Semiconductor[u]											x	x	x					
Continental Device[v]												x	x	x	x	x	x	
International Business Machines[w]												o	o	o	o	o	o	o
International Telephone and Telegraph[x]												o	o	o	o	x	x	x
Teledyne[y]												x	x	x	x	x	x	x
Siliconix[z]													x	x	x	x	x	x
Union Carbide Electronics[aa]														o	x	x	x	x

Table 4-1 (*continued*)

Transistor firm	*1951*	*1952*	*1953*	*1954*	*1955*	*1956*	*1957*	*1958*	*1959*	*1960*	*1961*	*1962*	*1963*	*1964*	*1965*	*1966*	*1967*	*1968*
Dickson Electronics[bb]															x	x	x	x
KMC Semiconductor[cc]															x	x	x	x
Solitron															o	x	x	x
Number of other transistor firms listed in directory[dd]	n.a.	0	4	5	7	6	6	6	8	17	21	11	18	22	23	21	26	26
Total number of transistor firms listed in directory[dd]	n.a.	5	15	18	26	26	28	28	34	47	53	42	48	51	50	50	53	52

Sources: *Electronics Buyers' Guide* (McGraw-Hill, annual; referred to below as the directory); other directories; various articles in the trade press; correspondence with firms.

n.a. Not available.

a. The years a firm is listed in the *Electronics Buyers' Guide* as selling transistors are denoted by "x." Other years a firm is believed to have commercially produced transistors are denoted by "o."

b. Western Electric produces transistors only for internal use and the government market and so is seldom listed in the directory.

c. By 1956, Bogue controlled Germanium Products and Radio Development and Research.

d. Clevite acquired a majority interest in Transistor Products in 1953 and total ownership in 1955. ITT, another transistor producer, purchased the Clevite semiconductor operations in 1965.

e. Until about 1958, Motorola's transistor production was only for in-house needs and quite limited.

f. CBS sold its transistor plant and equipment to Raytheon, another semiconductor producer, in 1961. It reentered the semiconductor market in 1964 with a line of integrated circuits.

g. Apparently Radio Receptor left the transistor business after its transistor personnel departed about 1955 to start General Transistor.

h. Tung-Sol discontinued transistor fabrication around 1963, retaining only silicon rectifiers as semiconductor products.

i. Amperex is a subsidiary of Philips, the large Dutch electrical firm.

j. Honeywell sold its semiconductor division to Solitron Devices in 1965, apparently because it was unprofitable.

k. Ford acquired Philco in 1961. Philco-Ford announced it would stop selling transistors in 1963 in order to concentrate on integrated circuits. However, with the acquisition of General Micro-Electronics in 1966, it reentered the transistor business on a limited scale.

l. Pacific Semiconductors, an early spin-off from Hughes was established by Thompson Ramo Wooldridge in 1954. Its name was later changed to TRW Semiconductors.

m. General Transistor merged into General Instrument in 1960.

n. Delco Radio is a division of the General Motors Corporation.

o. Sperry Rand had sold certain of its transistor operations to Solitron Devices in 1967 and discontinued the rest in 1968.

p. Rheem Semiconductor, a 1959 spin-off from Fairchild, was taken over by Raytheon, another semiconductor firm, in 1961.

q. A new group of investors took over National Semiconductor in 1967, and began what was really a new firm within the remaining corporate structure of National Semiconductor.

r. Solid State Products was acquired by Unitrode in mid-1967.

s. Crystalonics, a spin-off from Transitron, was acquired by Teledyne in 1965.

t. Former employees of General Transistor and RCA founded Silicon Transistor about 1960.

u. Micro Semiconductor was established by former Thompson Ramo Wooldridge employees in the early 1960s.

v. Continental Device, a spin-off from Hughes, was acquired by Teledyne in 1967.

w. In 1962, IBM began producing silicon transistors for the hybrid circuits used in its 360-series computers. It had, however, developed the equipment and production ability to produce germanium transistors in 1957. Much of this capability was then transferred to Texas Instruments for large-scale production. Since IBM produces transistors only for in-house use, it is not listed in the directory.

x. ITT began producing transistors in the United States in 1962 when the company acquired National Transistor. Just when National Transistor, a spin-off from Clevite, first produced transistors is not known.

y. Amelco was founded in 1962 by former Fairchild employees as a Teledyne subsidiary.

z. Siliconix was founded in 1962 by former employees of Westinghouse and Texas Instruments.

aa. Union Carbide Electronics, founded by former Amelco employees, began semiconductor production in 1964.

bb. Dickson Electronics was founded early in the 1960s by former Motorola personnel.

cc. Former RCA employees established KMC Semiconductor about 1963.

dd. These figures indicate the number of (other) firms listed in the *Electronics Buyers' Guide* as selling transistors and may deviate from the actual number. See text for further discussion.

the firms found in the table are listed in the *Electronics Buyers' Guide*[3] as selling transistors for at least three years during the 1952–68 period. The specific years a firm appears in the directory are indicated by the letter "x" in the table.

The number of transistor firms found in the *Guide* but not identified in the table is shown at the bottom of the table, along with the total number of transistor firms appearing in the directory. On balance, the number of firms listed in the directory probably overestimates the actual number of transistor producers, though apparently not by much.[4] A second caveat concerns production dates. A company may not be listed in the *Guide* for a year or two, or even longer, after initiating production. Also, a firm may be dropped from the listing for several years for no apparent reason. Those years when evidence from other directories or the trade literature indicates a firm was producing transistors even though it did not appear

3. *Electronics Buyers' Guide* (McGraw-Hill, annual). There are other directories listing firms selling transistors and other semiconductor devices, such as the *Electronic Industries Buying Guide* (Philadelphia: Chilton, annual). Primary reliance was placed on the *Electronics Buyers' Guide* because it lists transistor firms over the longest period and because its coverage of companies is comprehensive.

The listings of various directories differ considerably with regard to firms that have recently entered or plan shortly to enter the semiconductor business. Generally, however, they agree on the major producers.

4. The figure from the directory may deviate from the actual figure for several reasons. First, the directory may omit some of the many minor firms that produced transistors for only a year or two and then went out of business. Second, it may include some companies that intended to enter transistor production, but then for technical or other reasons never did. Third, it generally excludes firms like IBM that produce only for in-house use, since it is primarily concerned with firms selling transistors. Fourth, it lists a few firms that are not transistor producers but rather importers or wholesalers. (It also lists foreign firms, but they were not included in Table 4-1.) Finally, it occasionally gives separate listings for different divisions or plants of the same firm.

The Business and Defense Services Administration (BDSA) of the U.S. Department of Commerce estimates that there were thirty American firms producing transistors in 1959, as against the thirty-four firms listed in the directory. (See BDSA, *Electronic Components: Production and Related Data, 1952–1959* [1960], Table 7.) This source also estimates that there were sixty firms producing semiconductor devices of all types in 1959. Elsewhere BDSA indicates that the number of semiconductor firms was sixty-seven in 1958, eighty-three in 1962, and one hundred in 1965 (BDSA, "Semiconductor Growth Study," ca. 1966; processed). These estimates are all substantially smaller than the number of firms listed in the *Electronics Buyers' Guide* as selling some type of semiconductor. Consequently, no attempt was made to estimate the total number of semiconductor firms with the procedure used here for transistor firms.

in the *Guide* are shown in Table 4-1 by the letter "o." Since the evidence is sometimes conflicting, the dates shown should be considered estimates.

Changes in company ownership and name, which frequently go together, created another problem in constructing Table 4-1. General Transistor, for example, merged into General Instrument in 1960 and thereafter sold transistors under the latter name, but the change marked neither the departure of an independent transistor company from the industry nor the birth of a new one. Such changes occur only when two transistor firms merge or when a firm fails and its assets are acquired by another firm. Except for such cases, wherever the necessary information is available Table 4-1 indicates as single entities companies that have changed names or ownership. Earlier names and changes in ownership are given in the table notes.

Despite these problems and limitations, Table 4-1 does illustrate several important features in the historical development of the American semiconductor industry. First, receiving tube producers were among the first to jump into transistor production. General Electric, Raytheon, and RCA began in 1951, the first year the transistor was commercially produced, and by 1954, all eight receiving tube producers were manufacturing transistors. Second, new firms were slower to respond to the new technology emanating from Bell Laboratories, but still entered in large numbers within five years after commercial production was begun by Western Electric. Third, entry into the industry has been easy. Many firms entered the industry during the fifties and sixties, though entry apparently slackened during the latter decade. Finally, while entry has been easy, survival has been difficult. Over a quarter of the companies listed in Table 4-1 have either abandoned the transistor business or been absorbed by other transistor firms. Many more casualties lie buried among the number of "other transistor firms" noted at the bottom of the table—according to the *Guide,* the firms in this category lasted on average only two years in the transistor business.

Contributions by Firms

Many different kinds of company activities have produced the rapid progress in semiconductor technology and thus promoted the industry's swift growth. As pointed out in Chapter 1, these activities contribute either to the innovative process or the diffusion process.

Innovative Process

The most widely used measures of companies' innovative activity are patents, major innovations, and R&D expenditures. Each, however, has certain drawbacks.[5] Patents differ greatly in technical and economic importance. Moreover, they are not acquired on all technological advances, and the propensity to patent may vary over time and between firms. Lists of major innovations are always arbitrary to some extent. And there is some evidence to suggest that less spectacular, follow-on innovations may be more important in the aggregate than major innovations.[6] Research and development expenditures measure inputs to the innovative process and may not closely reflect outputs.

Since the limitations of these three measures differ, they are complementary and serve as useful checks on each other.[7] We therefore consider all three in assessing company efforts in the innovative process.

PATENTS. The number of semiconductor patents awarded to Bell Laboratories (and Western Electric), to the eight receiving tube producers, and to the thirteen new firms holding fifty or more patents are shown in Table 4-2 for the years 1952–68. The table also indicates the total number of semiconductor patents issued to seventeen other new firms that individually received less than fifty patents.

5. For a more extensive examination of the strengths and weaknesses of these three measures, see Simon Kuznets, "Inventive Activity: Problems of Definition and Measurement," in *The Rate and Direction of Inventive Activity: Economic and Social Factors,* A Conference of the Universities-National Bureau Committee for Economic Research and the Committee on Economic Growth of the Social Science Research Council (Princeton University Press for the National Bureau of Economic Research, 1962), pp. 19–43; Dennis C. Mueller, "Patents, Research and Development, and the Measurement of Inventive Activity," *Journal of Industrial Economics,* Vol. 15 (November 1966), pp. 26–37; and Jacob Schmookler, *Invention and Economic Growth* (Harvard University Press, 1966), Chap. 2.

6. See John L. Enos, "Invention and Innovation in the Petroleum Refining Industry," in *The Rate and Direction of Inventive Activity,* pp. 299–321; and Samuel Hollander, *The Sources of Increased Efficiency: A Study of Du Pont Rayon Plants* (M.I.T. Press, 1965), Chap. 4.

7. In addition, they may reflect the level of activity at various stages of the innovative process. A popular image of the innovative spectrum portrays R&D expenditures producing inventions (and patents), which in turn lead to innovations. In practice, the sequence is not so clear-cut. R&D expenditures often occur after an invention and even after an innovation. In addition, the technological advance covered by a patent is often in commercial use long before the patent is awarded.

Table 4-2. Semiconductor Patents Awarded to Firms in the United States, 1952–68

Type and name of firm	*1952*	*1953*	*1954*	*1955*	*1956*	*1957*	*1958*	*1959*	*1960*	*1961*	*1962*	*1963*	*1964*	*1965*	*1966*	*1967*	*1968*	*1952–68*
Bell Laboratories[a]	34	47	36	27	48	34	51	72	67	67	55	54	41	64	59	41	38	835
Receiving tube firms																		
RCA	14	19	16	10	44	45	57	46	34	33	33	27	41	58	75	63	53	668
General Electric	5	7	6	11	25	13	35	34	26	36	42	35	38	86	76	64	41	580
Westinghouse	2	6	2	4	9	7	21	25	30	21	23	18	23	78	54	46	41	410
Sylvania	1	4	6	1	11	14	19	9	13	7	13	10	9	14	7	12	8	158
Philco-Ford	0	0	0	1	3	4	7	13	14	13	14	9	8	12	17	9	6	130
Raytheon	0	1	0	1	7	3	10	8	6	6	4	7	3	5	6	4	1	72
Tung-Sol	0	0	0	1	0	0	3	1	0	2	5	1	2	5	2	4	0	26
Columbia Broadcasting System	0	0	0	2	2	0	1	1	0	0	1	0	0	1	1	0	0	9
Subtotal	**22**	**37**	**30**	**31**	**101**	**86**	**153**	**137**	**123**	**118**	**135**	**107**	**124**	**259**	**238**	**202**	**150**	**2,053**
New firms																		
International Business Machines	1	1	0	1	4	6	15	22	25	37	96	53	47	56	70	46	41	521
Texas Instruments	0	0	0	0	0	2	9	8	22	13	20	24	16	43	35	52	42	286
Motorola	0	1	3	2	5	11	8	8	9	10	10	6	8	26	25	39	19	190
Hughes	0	0	5	3	9	12	18	11	3	9	10	7	9	19	16	16	13	160
Honeywell	0	0	0	0	1	2	5	17	13	12	38	16	15	11	18	9	3	160
Sperry Rand	0	0	0	1	0	5	5	7	7	6	13	9	11	35	18	15	7	139
General Motors	0	0	0	0	0	0	6	5	8	12	3	5	8	26	31	16	13	133
International Telephone and Telegraph[b]	2	2	0	4	4	2	11	15	10	6	18	4	6	7	5	9	6	111
Clevite	0	0	3	3	2	1	3	3	7	7	4	12	17	10	3	2	1	78
Bendix	0	1	1	0	0	3	2	4	7	9	0	5	3	15	16	6	5	77
Thompson Ramo Wooldridge	0	0	0	0	0	1	3	7	3	18	11	11	2	4	0	0	0	60
Fairchild	0	0	0	0	0	0	0	0	1	4	2	3	6	10	10	11	5	52
Sprague	0	0	0	0	2	2	2	8	2	1	4	3	5	6	6	5	6	52
17 other new firms	1	3	1	1	10	7	16	22	15	12	21	9	7	30	33	10	23	221
Subtotal[c]	**4**	**8**	**13**	**15**	**37**	**54**	**103**	**137**	**132**	**156**	**250**	**167**	**160**	**298**	**286**	**236**	**184**	**2,240**
Total[c]	**60**	**92**	**79**	**73**	**186**	**174**	**307**	**346**	**322**	**341**	**440**	**328**	**325**	**621**	**583**	**479**	**372**	**5,128**
Percentage of total																		
Bell Laboratories	*56*	*51*	*46*	*37*	*26*	*20*	*17*	*20*	*21*	*20*	*12*	*16*	*13*	*10*	*10*	*9*	*10*	*16*
Receiving tube firms	*37*	*40*	*38*	*42*	*54*	*49*	*50*	*40*	*38*	*34*	*31*	*33*	*38*	*42*	*41*	*42*	*40*	*40*
New firms	*7*	*9*	*16*	*21*	*20*	*31*	*33*	*40*	*41*	*46*	*57*	*51*	*49*	*48*	*49*	*49*	*50*	*44*

Source: U.S. Patent Office records.

a. Includes patents awarded to Western Electric.

b. Excludes patents awarded to the International Standard Electric Corporation, a wholly owned ITT subsidiary that serves as a holding company for many of ITT's foreign subsidiaries.

c. Includes patents awarded to the major semiconductor producers, but not to nonproducers or many of the small new firms. Consequently, the total does not include all semiconductor patent awards.

These patents cover new semiconductor devices, new methods of semiconductor fabrication, new equipment for manufacturing and testing semiconductors, and new applications of semiconductors in final electronic products where their use is important enough to have merited explicit notation in the title of the patent. A sample of patent titles counted in the table includes the following: tunnel diode, crystal diode encapsulation, method of passivating semiconductors, testing apparatus for crystal diodes, method of growing germanium crystals, and transistorized ignition system.

In using patents to estimate innovative output, the lag between the date a firm applies for a patent and the date it is granted must be taken into account. Delays can be as long as ten years and shorter than one, with an average pendency period of between three and four years. Consequently, the patents shown in Table 4-2 reflect innovative output for a period some three to four years earlier than the year they were issued.

With this consideration in mind, what do the patent data indicate about the innovative process and the contributions ot the various types of firms to that process? First, over the entire seventeen-year period, Bell Laboratories acquired more semiconductor patents than any other firm, accounting for 16 percent of all the patents shown in the table. Receiving tube producers are the next largest patent holders—RCA is second, General Electric third, and Westinghouse fifth. As a bloc, these eight firms account for 40 percent of the total number of patents. All the firms have obtained some patents, and all but CBS and Tung-Sol have more than fifty.

The new firms as a group account for 44 percent of the total patents, a larger share than either Bell Laboratories or the receiving tube firms, but this only reflects their greater number. Individually, many new firms have fewer than fifty semiconductor patents, and some have none. The largest patent holder in this group, and the fourth largest overall, is IBM. Most of its 521 patents cover new applications for semiconductors, particularly in data-processing equipment. If such patents were excluded, IBM's rank on the patent ladder would drop considerably.[8] After IBM,

8. Patents excluding applications of semiconductors were collected for 1962–64. Although the total number of patents for these years fell by 48 percent, the relative positions of most firms stayed about the same. Only three firms changed more than four places in this three-year period: IBM fell from first to eighth in the list of thirty-nine firms, Sperry Rand from eleventh to twenty-third, and ITT from twelfth to twentieth.

Texas Instruments is the largest patent holder among the new firms and Motorola is next.

Considering only the total number of patents hides interesting trends. The level of patents has changed over time and with it the relative contributions of firms. During the 1950s, the annual number of patents awarded rose markedly, though the process was sporadic. Large jumps were often followed by slight declines for a year or two before another jump occurred. The same instability marked the 1960s, but during this period the upward trend tapered off.

Bell Laboratories accounted for roughly half of all semiconductor patents early in the 1950s. As the industry expanded during the second half of the decade, Bell's share fell to about 20 percent even though it increased its own pace of patent acquisition. In the 1960s, Bell's patent awards fluctuated from year to year, but no long-run increase was apparent and its share of the total declined to around 10 percent.

Except for CBS, Tung-Sol, and Philco, the receiving tube producers were actively acquiring semiconductor patents in the early 1950s and even the three exceptions had entered the competition by 1955. (This supports the evidence noted earlier indicating that the receiving tube firms were quick to realize the importance of the new semiconductor technology.) Despite their early efforts, the number of patents awarded to the receiving tube firms remained small until 1956. The large jump in both the number and share of their patents in that year reflects an increase in their innovative activities some four years earlier, about the time many of these companies were trying to introduce transistors into the market. Also stimulating their R&D efforts at this time was the 1952 symposium (which is considered in more detail later) at which Bell Laboratories divulged much of its semiconductor technology. During the second half of the 1950s, the receiving tube producers accounted for nearly 50 percent of all semiconductor patents awarded. This percentage fell in the 1960s, varying between 31 and 42 percent.

The new firms started later in the patent-acquisition game, which in part explains the relatively small percentage that have accumulated fifty or more patents. Even by 1958 these firms as a group accounted for only 33 percent of the total, but they have since gradually increased their share and now obtain about half of all patents awarded annually.

MAJOR INNOVATIONS. The major process and product innovations in the semiconductor industry were identified and briefly described in Tables

Table 4-3. Major Semiconductor Innovations in the United States, by Firm, 1951–68[a]

	All innovations[b]		*Process innovations*		*Product innovations*	
Type and name of firm	*Number*	*Percent*	*Number*	*Percent*	*Number*	*Percent*
Bell Laboratories and Western Electric	**5.5**	*36*	**5.0**	56	**3.5**	29
Receiving tube firms						
General Electric	2.5	*17*	2.0	*22*	1.5	*13*
Philco-Ford	1.0	*7*	1.0	*11*	1.0	*8*
RCA	0.5	*3*	0.0	*0*	0.5	*4*
Subtotal	**4.0**	*27*	**3.0**	*33*	**3.0**	25
New firms						
Texas Instruments	2.0	*13*	0.0	0	2.0	*17*
Fairchild	2.5	*17*	1.0	*11*	2.5	*21*
International Business Machines	1.0	*7*	0.0	*0*	1.0	*8*
Subtotal	**5.5**	*37*	**1.0**	*11*	**5.5**	*46*
Total	**15.0**	***100***	9.0	***100***	**12.0**	***100***

Sources: Tables 2-1 and 2-2.

a. The tunnel diode and 3-5 compounds are excluded since these major innovations are attributed to foreign firms. Where two firms are responsible for introducing a major innovation, each receives half the credit.

b. When major new process innovations led directly to a new product innovation, the two were counted as only one innovation in this column. This explains why the sum of the product and process innovations exceeds fifteen, the total noted for all major innovations.

2-1 and 2-2. The number of these innovations introduced by specific American firms and the various types of companies are shown in Table 4-3.

Bell Laboratories has produced 5½ major innovations, which is more than twice the number introduced by any other American firm. General Electric and Fairchild, each with 2½ innovations, share second place.

Bell Laboratories has particularly dominated the development of new process innovations, generating 56 percent of the total. The only other firms to produce major process innovations are General Electric, Philco-Ford, and Fairchild. Credits for product innovations are more widely dispersed. Bell Laboratories accounts for more than any other firm, but its total is surpassed by the new firms as a group.

The six firms with the greatest number of patents accounted for 75 percent of the major innovations. While this suggests that activities generating major innovations also generate patents and vice versa, the association between the two should not be exaggerated. Westinghouse has acquired a large portfolio of patents but introduced no major innovations, while the reverse is true of Fairchild and Philco-Ford. Moreover, the number of major innovations introduced by the top six does not appear closely related to the number of patents they have accumulated.[9] To be sure, Bell Laboratories has produced more patents and more major innovations than any other firm, but its share of the total major innovations is double its share of total patents. Moreover, the second largest patent holder, RCA, has only half a major innovation to its credit, while Texas Instruments with the fewest patents has introduced two major innovations.

R&D EXPENDITURES. Expenditures for R&D in the research-intensive semiconductor industry are high. In 1965, they exceeded 6 percent of sales. Earlier in the industry's history, when sales were smaller, this percentage was much higher, nearly 18 percent in 1959 and 27 percent in 1958.[10] Although, as discussed below, the government has financed a good part of this R&D effort, semiconductor firms have spent sizable amounts of their own funds on R&D.

Individual company efforts are difficult to assess and compare because semiconductor R&D expenditures broken down by firm are seldom available. Only a few release their total R&D expenditures, and these are generally the large diversified firms whose semiconductor R&D is only a small unknown part of their overall R&D effort. As a result, the desired

9. The rank correlation coefficient between the patents and major innovations of these firms is only 0.34, which is not significantly (at the 95 percent level) different from zero. In part this result may reflect the fact that the patent data include applications of semiconductors in final electronic equipment while the major innovations do not; however, as noted earlier, the relative ranks of most firms are believed to remain about the same when application patents are excluded. One of the exceptions noted was IBM, but any change in its patent rank would probably reduce the rank coefficient, since IBM is fourth among the six firms in both patents and major innovations.

10. These figures are based on R&D expenditures reported in BDSA, *Semiconductors: U.S. Production and Trade* (1961), Table 9.

The 6 percent figure for 1965 is found in Organisation for Economic Co-operation and Development, *Electronic Components: Gaps in Technology* (Paris: OECD, 1968), p. 179, and is based on a BDSA reply to an OECD inquiry. However, OECD also indicates that the number of scientists and engineers in the American semiconductor industry increased by 18 percent between 1959 and 1965. If R&D expenditures increased by a similar amount, they would have totaled $85 million, or nearly 10 percent of semiconductor sales in 1965. It appears therefore that the 6 percent figure is too low and that the true figure is between 6 and 10 percent.

Table 4-4. Semiconductor Research and Development Expenditures in the United States, 1959, and Patents Acquired, 1962–64, by Type of Firm

Type of firm	1959 R&D expenditures: Millions of dollars	1959 R&D expenditures: Percent	Patents acquired, 1962–64[a]: Average annual number	Patents acquired, 1962–64[a]: Percent	Patents per $1 million of R&D expenditures (col. 3÷col. 1)
Bell Laboratories and receiving tube firms[b]	39.9	*57*	96	*54*	2.4
New firms[c]	30.3	*43*	81	*46*	2.7
Total	70.2	*100*	177	*100*	...

Sources: U.S. Department of Defense, Survey of 64 Semiconductor Companies, 1960, unpublished tabulations; and U.S. Patent Office.

a. Patents cover new semiconductor devices, new methods of semiconductor fabrication, and new manufacturing and testing equipment for semiconductors, but not new applications of semiconductors in final electronic products since expenditures on new applications were not included in the R&D data.

The average lag of four years between R&D expenditures and patents allows time for R&D to produce patentable results and for patent applications to be processed. Since the latter delay alone averages between three and four years, it may appear that this lag is too short. However, when the relationship between R&D expenditures and patents is estimated by regression analysis, the coefficient of determination (R^2) is higher when 1963 patents are used than when either 1962 or 1964 patents are used. This suggests that the average lag of four years is appropriate.

The average number of patents acquired annually over the 1962–64 period is considered rather than the number for 1963 alone, because R&D expenditures in any one year may generate patents over several years and because companies' patent acquisitions often fluctuate considerably from one year to the next.

b. R&D figures for Bell Laboratories are combined with those for the receiving tube firms to avoid disclosing figures of individual firms.

c. Thirty new firms are considered; this includes all the significant semiconductor firms (except for Bell Laboratories and the receiving tube firms) in 1959. These thirty do not correspond exactly to the thirty new firms considered in Table 4-2, though the overlap is great.

R&D data are available only for 1959, a limited but formative period in the young industry's development.[11]

Table 4-4 shows that in that year Bell Laboratories and the eight receiving tube producers spent 57 percent and the new firms 43 percent, respectively, of the R&D funds in the semiconductor field. The table also shows that the first group of firms accounted for a slightly greater share of 1959 R&D expenditures than of semiconductor patents acquired during the 1962–64 period, indicating that the new firms obtained slightly more patents for each million dollars spent on R&D. These differences, however, are not great, and when individual firms, rather than groups of firms, are

11. The data were collected by an ad hoc Department of Defense survey of the semiconductor industry in 1960. They were made available on the understanding that individual company figures would not be disclosed. The definitions of R&D activities used in the survey are those used by the National Science Foundation.

compared, the number of semiconductor patents a firm receives is found to be closely related to the amount it spends on R&D.[12]

Thus, the data on patents, major innovations, and R&D expenditures indicate that AT&T (including Bell Laboratories and Western Electric) has contributed more to the innovative process and semiconductor technology than any other firm. The number of its patents and of its major innovations greatly exceeds the number of each attributed to other firms, and the amount it spends on R&D in the semiconductor field represents a sizable part of all such expenditures.

The receiving tube firms have also made a large contribution toward advancing semiconductor technology. Though they represent only a small portion of all semiconductor firms, they account for 40 percent of the patents, 27 percent of the major innovations, and along with Bell Laboratories over half the R&D expenditures on semiconductor activities in 1959, and probably other years as well. General Electric, RCA, Westinghouse, and Philco-Ford are the most important contributors in this group. Only CBS and Tung-Sol appear to have added little.

The many new firms together are responsible for 44 percent of semiconductor patents, 37 percent of major innovations, and (in 1959) 43 percent of R&D expenditures. In general, their individual contributions to the innovative process are small, though there are exceptions such as Texas Instruments, IBM, Motorola, and Fairchild. But even these leaders among the new companies seldom rank among the very top firms in patents, major innovations, and R&D expenditures.[13]

12. By using regression analysis on data for thirty-eight firms, the following relation was obtained between the amount of semiconductor R&D conducted by a firm in 1959 (E_i) and the average annual number of semiconductor patents (excluding application patents) it acquired during the 1962–64 period (P_i):

$$P_i = \underset{(0.54)}{0.59} + \underset{(0.17)}{2.21}\, E_i$$

The standard errors shown in parentheses indicate that R&D expenditures are significantly related at over the 99 percent probability level to the number of patents a firm obtains. The coefficient of determination (R^2) is 0.82, indicating that R&D expenditures account for 82 percent of the variation among firms in semiconductor patent awards.

13. Although the contributions of new firms to the innovative process are relatively small, we cannot conclude that a market structure with fewer new firms and more firms like Bell Laboratories and the receiving tube producers would necessarily promote more innovative activity. Before such a conclusion is possible, the

Diffusion Process

The level of technology a firm uses in production is the crucial determinant of its ability to compete in the semiconductor market. Any advantage it may enjoy from other factors is quickly lost if it falls behind in introducing and using new developments in its production lines. This section, therefore, examines market shares to assess the contributions of various companies and types of companies in the diffusion process.

The methodology employed here, and in the two following chapters, to measure company contributions to the diffusion process rests on an important assumption that should be spelled out in more detail. Specifically, it is assumed that the market share of a semiconductor firm depends on the technology it uses, which in turn is a function of the company's contributions to the innovative and diffusion processes, and that in comparison all other factors affecting market shares have a negligible effect. For most industries, this is not a valid assumption, since market shares are significantly affected by a host of other factors—advertising, pricing and profit policies, vertical integration and control over strategic factor inputs or market outlets, buyer-seller ties, monopsonistic power, and so on. Yet for the semiconductor industry the assumption does seem reasonable. As documented in Chapter 2, technological change has proceeded at an extraordinary pace in this industry. As a result, a firm with the latest technology may have production costs 20 percent, 40 percent, or more below the production costs of firms using technology that has been obsolete for six months or a year. Moreover, the company with superior technology usually produces devices with greater reliability and capabilities. Thus it is not surprising that the effects of other factors on a firm's competitive position and share of the market appear insignificant by comparison.

Given that market shares depend on innovative and diffusion performance, company contributions to the diffusion process can be assessed

---

innovative performance of the various types of firms must be adjusted for differences in their market shares. If, for example, new firms accounted for around 40 percent of the innovative activity in the industry but had only 20 percent of the market, their innovative effort would not appear deficient. (The next section, however, shows that the new firms have acquired larger market shares than the other types of firms.) In addition, other factors such as the effect of easy entry conditions and of new firms on the innovative activity of the receiving tube firms and Bell Laboratories should be taken into account.

by comparing data on market shares and innovative performance. If firms capture a share of the market exceeding their innovative contributions, as new firms have in the American market, this implies that they have been particularly instrumental in diffusing new semiconductor technology.[14]

Market shares of the major semiconductor producers at three-year intervals over the 1957–66 period are shown in Table 4-5.[15] Information for earlier years is sparse, though it is known that Hughes, which was a major manufacturer of diodes and rectifiers, led the industry in semiconductor sales early in the 1950s. By 1957, Hughes had dropped to third place, and its relative importance continued to decline thereafter. By 1961 the firm was no longer among the top ten producers.

Raytheon, a receiving tube firm, was the major producer of transistors for the few years around 1952 when hearing aids provided the principal transistor market. The second largest seller of transistors at this time was a new firm, Germanium Products, which soon merged into Bogue Electric.

In 1954, Texas Instruments, building on research at Bell Laboratories, introduced the first silicon transistor. The use of silicon rather than germanium increased the temperature and frequency ranges over which the transistor could operate, thereby opening up a vast new market, particularly in products for military use. Texas Instruments' lead in this development lasted some two or three years, a remarkably long time in an in-

14. Alternatively, one might assess a firm's contribution to the diffusion process directly from its market share, for a firm with a large market share must be relatively fast and effective at introducing and using new semiconductor technology in production. This approach, however, ignores the fact that the firm developing a new semiconductor innovation is in a more favorable position than other firms to introduce the innovation into its production operations and to exploit it commercially. Thus a firm may acquire a large share of the market by superior performance in the innovative process even though it is not particularly adept at diffusion.

15. Firms do not release to the public information on their production, shipments, or sales of semiconductors. Shipments may differ from production since they exclude net increases in inventory holdings as well as devices that do not meet minimum specifications and must be rejected. Shipments may differ from sales since one sale can call for shipments over a period of several years. Nevertheless, production and sales are often used as synonyms for shipments, as is done throughout this study. Firms do report their shipments to the Business and Defense Services Administration and the Electronic Industries Association. These organizations publish industry totals for various types of semiconductors, but carefully guard the figures of individual firms. Consequently, the market shares shown here are estimates based on the sources noted in the table.

Table 4-5. U.S. Semiconductor Market Shares of the Major Firms, Selected Years, 1957–66[a]

Type and name of firm	Percentage of market			
	1957	*1960*	*1963*	*1966*
Western Electric	**5**	**5**	**7**	**9**
Receiving tube firms				
General Electric	9	8	8	8
RCA	6	7	5	7
Raytheon	5	4	b	b
Sylvania	4	3	b	b
Philco-Ford	3	6	4	3
Westinghouse	2	6	4	5
Others	2	1	4	3
Subtotal	**31**	**35**	**25**	**26**
New firms				
Texas Instruments	20	20	18	17
Transitron	12	9	3	3
Hughes	11	5	b	b
Motorola	b	5	10	12
Fairchild	b	5	9	13
Thompson Ramo Wooldridge	b	b	4	b
General Instrument	b	b	b	4
Delco Radio	b	b	b	4
Others	21	16	24	12
Subtotal	**64**	**60**	**68**	**65**
Total	**100**	**100**	**100**	**100**

Sources: H. Gunther Rudenberg, *The Outlook for Integrated Circuits* (Cambridge, Mass.: Arthur D. Little, Inc., 1967); Arthur D. Little, Inc., *Trends in Integrated Circuits and Microelectronics* (ADL, 1965); Arthur D. Little, Inc., *Seven Western Semiconductor Manufacturers* (ADL, 1961); Herbert S. Kleiman, "The Integrated Circuit: A Case Study of Product Innovation in the Electronics Industry" (D.B.A. thesis, George Washington University, 1966), Chap. 5; Barry Miller, "Competition to Tighten in Semiconductors," *Aviation Week and Space Technology* (Sept. 23, 1963), pp. 56–68; and other articles from the trade press.

a. Market shares are based on company shipments and include in-house and government sales.

b. Not one of the top semiconductor firms for this year. Its market share is included under "others."

dustry where firms usually can duplicate a new device and second-source the innovator within six months. The silicon transistor generated considerable profits for Texas Instruments and catapulted the small firm to the top position in semiconductor sales.

The second largest semiconductor producer in the late 1950s was Transitron, a small new firm whose growth and success were also built on a new product. Bell Laboratories developed the first gold-bonded diode, but Transitron was the first to work out the many problems associated with large-volume production and to achieve yields high enough to permit a price competitive with the less reliable point-contact diode then in use. The gold-bonded diode launched the new firm and helped support subsequent efforts on other semiconductor devices. But in the early 1960s profound changes buffeted the semiconductor industry. A wave of new technology using silicon and relying on the oxide masking, diffusion, planar, and epitaxial techniques arose, and prices fell rapidly, stymieing the dollar value of semiconductor sales after nearly a decade of rapid growth. In these circumstances, many firms suffered, and Transitron particularly so. The company fell from second place with about 10 percent of the semiconductor market in 1960 to eleventh place with only 3 percent of the market in 1963. Contributing to Transitron's troubles was its relatively low R&D effort;[16] over the 1952–68 period it acquired only 26 semiconductor patents (according to the data collected for Table 4-2) while Texas Instruments, for example, accumulated 286.

Two other new firms—Motorola and Fairchild—have climbed to top positions on the industry's sales ladder. Both are relative latecomers. Fairchild spun off from Shockley Laboratories in 1957. Motorola's semiconductor production was small and primarily for its own consumption until 1958. The two firms have ridden the wave of new semiconductor technology that apparently capsized Transitron. Fairchild controls the strategic patents on the planar process, and Fairchild, Motorola, and Texas Instruments have extensively used the new technology to batch-produce silicon diodes, transistors, and integrated circuits. Among the new firms that have acquired very large market shares, Motorola is the only one that was not a small company at the time it began manufacturing semiconductors.

Receiving tube firms have occupied the second echelon in the ranks of semiconductor producers. In 1957, six receiving tube firms and Western Electric followed Texas Instruments, Transitron, and Hughes as the industry's largest producers. Tung-Sol and CBS are the only receiving tube firms that do not at some time appear among the ten largest semiconductor producers. General Electric and RCA each managed to hold between 5 and 10 percent of the total market over the entire period covered by the table. Raytheon, despite its early lead in transistors, and Sylvania did not remain among the top firms after 1960.

The fortunes of Philco-Ford and Westinghouse, the two other receiv-

16. See William B. Harris, "The Company That Started with a Gold Whisker," *Fortune* (August 1959), p. 142.

ing tube firms, have fluctuated. Philco enjoyed considerable success with its surface-barrier transistor and jet-etching techniques during the second half of the 1950s but, like Transitron, did not keep pace with the new technologies of the 1960s and suffered accordingly. In 1961, Ford acquired the floundering company, and several years later discontinued selling discrete semiconductor devices in order to concentrate on integrated circuits. Philco-Ford reentered this market in 1966 when it purchased General Micro-Electronics, a small new firm with expertise in MOS (metal oxide semiconductor) technology.

Westinghouse had a rather mediocre share of the semiconductor market in the 1950s. With substantial financial support from the Air Force, the company launched a major R&D effort to develop integrated circuits in the late 1950s and early 1960s.[17] Although the major breakthroughs that made the integrated circuit a reality came from Texas Instruments and Fairchild, the effort improved Westinghouse's market position.

Though the receiving tube firms have not matched the top two or three new firms in overall sales, they have outsold them in some product lines. General Electric and Westinghouse, for instance, have dominated the market for power devices, such as power rectifiers and silicon controlled rectifiers, reflecting their interest in heavy electrical equipment using such devices.

Western Electric supplied about 5 percent of the semiconductor market in 1957. Since then, its share has almost doubled. The firm's sales are restricted by the 1956 consent decree to AT&T sister companies and the government market. Its market share has grown because its sales of semiconductors and semiconductor equipment to Bell System telephone companies have expanded rapidly.[18]

Despite its growing market share, Western Electric, like the receiving

17. Initially, the more successful firms were reluctant to undertake research on such a radically new development. See Herbert S. Kleiman, "The Integrated Circuit: A Case Study of Product Innovation in the Electronics Industry" (D.B.A. dissertation, George Washington University, 1966), pp. 185–88.

18. In evidence supporting its request for a renewal of its British patent covering the transistor, Western Electric stated its sales of transistors and transistorized equipment to sister companies within AT&T grew from £0.1 million in 1957 to £46.6 million in 1963. Over the same period, sales for U.S. government business rose from a negligible amount to £2.8 million. See High Court of Justice, Chancery Division, "In the Matter of the Patents Act, 1949, and In the Matter of Letters Patent granted to Western Electric Company Incorporated . . . numbered 694,021 for Apparatus Employing Bodies of Semiconducting Material" (1964 W. No. 04; processed), Exhibit K.

tube firms, has trailed behind various new firms in sales volume over the past two decades. Indeed, the market shares enjoyed by a handful of new firms have greatly surpassed their contributions to the innovative process. Texas Instruments, by far the largest semiconductor vendor in 1957, had produced by that time a negligible portion of all semiconductor patents, and its share of major innovations did not come close to matching its market share. Other comparisons at various times reveal similar discrepancies for Transitron, Motorola, and Fairchild. Apparently these firms have been particularly adept in the diffusion process and have led in using new technology, developed in their own and other laboratories, to produce better and cheaper semiconductor devices. This achievement contributed not only to their own welfare but also to the growth and development of the American semiconductor industry in general. Adapting a new technological development for large-scale production is often more difficult and expensive than producing the initial working model in the laboratory.

The new firms with large market shares are not the only companies to spur the rapid diffusion of semiconductor technology. In a few areas such as power rectifiers, the receiving tube firms have led the way in using new technology, though this is primarily technology that they themselves developed. Also, many new companies that have never acquired a large share of all semiconductor sales have nevertheless stimulated the swift diffusion of certain new semiconductors. They are often early producers of such devices, challenging the leadership of the larger firms in product lines with great growth potential. The latter usually respond to the challenge and extend their leadership to these new products as well. Occasionally, though, a new firm will capture and hold a sizable share of the market for a new product. The impact that such new firms can have is illustrated by the role they played in assuring the rapid diffusion of integrated-circuit and MOS technology.

In 1967, six years after integrated circuits were first produced commercially, Fairchild, Texas Instruments, Motorola, Signetics, and Westinghouse (which has since dropped out of the market) were the major integrated-circuit producers, accounting for 80 percent or more of the market.[19] About twenty-five other firms, including some in-house producers, battled for the rest of the market. A number of these firms, like

19. Fairchild held 32 percent of the market, Texas Instruments 24 percent, Motorola 12 percent, and Signetics and Westinghouse between 6 and 12 percent each. See Rudenberg, *Outlook for Integrated Circuits,* Tables 7 and 8.

Amelco and Siliconix, were recent spin-offs from other semiconductor firms.

The many small integrated-circuit producers, though they divide only a small part of the market, pose a continual challenge to the major suppliers. The latter are well aware that should they fall behind in adopting new technology they would soon lose their share of the market to the aggressive newcomers. Signetics, a 1961 Fairchild spin-off now controlled by Corning Glass, was itself once such a newcomer.

The most numerous and usually the most aggressive attacks on the market positions of the major integrated-circuit producers come from the new firms. But even the receiving tube producers have fielded a few troops—General Electric, RCA, Philco-Ford, Raytheon, and Sylvania, in addition to Westinghouse. Integrated circuits are displacing discrete devices in many applications, which offers these firms the opportunity to increase their share of the semiconductor market and perhaps replace Texas Instruments, Motorola, and Fairchild as the industry's principal suppliers. So far, however, their competition has not seriously threatened the three leaders.

Fairchild introduced the first MOS transistor. But after encountering difficulties with it and losing many MOS specialists, this company along with the other industry leaders shunned the new MOS techniques. Several years later, one Fairchild manager suggested another reason for the firm's delay in exploiting the MOS field: since Fairchild had developed a large, successful business in the conventional bipolar technology, the incentive to push another was small.[20]

Though neglected by Fairchild and other major semiconductor firms, the MOS field was quickly entered by several new firms:[21] General Micro-Electronics, which was founded by Fairchild personnel in 1963 and acquired by Philco-Ford in 1966, General Instrument, which also acquired Fairchild employees, and American Micro-Systems, which split off from General Micro-Electronics when the latter was purchased by Philco-Ford. These firms, despite some difficulties in realizing profits, demonstrated the potential of the MOS process for transistors and integrated circuits, and within a few years, Texas Instruments, Fairchild, and the other major firms were offering MOS devices too.

20. "Solid State," *Electronics,* Vol. 40 (Oct. 2, 1967), p. 40.

21. See "Where the Action Is in Electronics," *Business Week* (Oct. 4, 1969), p. 90; and "An Integrated Circuit That Is Catching Up," *Business Week* (April 25, 1970), pp. 134–36.

Thus, all three types of firms have contributed to the vitality and growth of the American semiconductor industry. Bell Laboratories and Western Electric, though under restraints in production and marketing, have greatly advanced the industry's technology. Similarly, the receiving tube producers have made important contributions to the innovative process. In the marketplace, they have at times and in certain fields offered the major producers formidable and stimulating competition, though more often they leave the impression of responding to the competition.

Conversely, the new firms, particularly the small ones, have been most effective in the diffusion process. From this group, the industry's sales leaders have risen—companies that have quickly and successfully taken new technology from the laboratory and adapted it for large-scale production. This group has also provided an endless stream of new firms to challenge the successful ones. Though few of these new companies ever capture a major share of the market, their mere presence promotes rapid diffusion. For, should the successful firms fail to foresee new technological trends or falter in some other way, the aggressive newcomers either prod them into swift corrective action or replace them.

Factors Affecting Company Performance

That AT&T and the receiving tube firms have been the major contributors to the innovative process is not surprising. These large firms could afford to maintain a sizable R&D effort in the semiconductor field, and were motivated to do so for several reasons. First, the new technology and patents produced could be traded with other companies both in the United States and abroad for other technology. Second, their wide range of activities increased the probability that, should their R&D lead to important results with applications outside the semiconductor field, they would still be of use to one of their other operations. This particularly increased the incentive to pursue basic research. Third, an R&D effort provided an in-house technical capability that could keep these firms abreast of the latest semiconductor developments, facilitate the assimilation of new technology developed elsewhere, and provide a base on which a future large-scale production effort could be built. Thus, R&D was a form of insurance against being forced out of the industry by the aggressive newcomers. Finally, a large R&D effort and competence in an advanced technology field like semiconductors enhanced the image of these firms in the eyes of their customers and the public at large.

For the new firms struggling to get established, the incentives for extensive R&D were not nearly as great. Often they had neither the time nor the money for R&D unless it related directly to the product or products they hoped to market. In addition, many new firms spun off from established firms and in the process borrowed the technology and R&D results of their mother firms.

New firms cannot, however, live forever on borrowed technology, and eventually a substantial in-house R&D effort becomes essential. Moreover, as new firms prosper and grow, the incentive to conduct R&D and the ability to support it increase for many of the same reasons as apply to AT&T and the receiving tube firms. As new firms such as Texas Instruments become established, they tend to make a greater contribution to the innovative process.

At the same time, their contribution to the diffusion process may start to taper off. At least, the reasons for new firms' having in the past contributed more to the diffusion of new technology than AT&T and the receiving tube firms raise this possibility. First, new companies initially had no vested interests, intellectual or commercial, in the threatened technology to dampen their enthusiasm for exploiting new semiconductor technology. The receiving tube companies often aggravated the adverse effect of vested interests by placing their semiconductor operations in their tube divisions, where tube experts, not semiconductor specialists, were the principal decision makers.

Second, semiconductor production offered the small new firms unusual opportunities for high rates of profit and growth. The potential rewards for owners and managers were great, enticing many to take exceptional risks in pursuit of these rewards.[22] Commensurate rewards were not available to the owners or managers of AT&T and the receiving tube firms because semiconductor production could not similarly affect the overall growth and profit rate of these large diversified concerns. This disparity in opportunities encouraged many enterprising persons to leave the large companies and join small firms or establish their own.

Third, AT&T and the receiving tube firms may have been less con-

22. Texas Instruments provides an illustration. During 1952–55, when the company was developing the silicon transistor, it poured $3 million into plant and equipment and incurred a net loss of over $1 million on its semiconductor operations. This was a big gamble for a little firm whose net sales in 1952 were only $20.5 million and net profits after taxes only $0.9 million. See Patrick E. Haggerty, "Strategies, Tactics, and Research," *Research Management,* Vol. 9, No. 3 (1966), p. 151.

cerned about maximizing profits than they were about simply achieving an acceptable level of profits. This would have reduced the incentive to take large risks except when necessary to maintain that level.

Finally, antitrust considerations and the 1956 consent decree curtailed AT&T's semiconductor market and limited the extent to which the firm could diffuse the benefits of its new techniques by using them to produce cheaper and better semiconductors. Antitrust considerations may have influenced the large receiving tube firms as well. Whenever antitrust activities are a concern, the incentive to dominate an industry must wane.

While the above factors can explain the incentives that motivated the new firms to make their greatest contribution to the diffusion process and the receiving tube firms and AT&T to the innovation process, they leave unanswered how new firms were able to enter a highly technical and research-intensive industry where a good portion of the technology was developed by AT&T and the receiving tube firms and to compete so successfully against the latter in the marketplace. Three principal reasons—easy accessibility to technology, the availability of venture capital coupled with insignificant scale economies, and government demand and financial support—are examined in the remaining sections of this chapter.

Availability of Technology

Access to state-of-the-art technology is essential if a new firm is to succeed in the semiconductor business. Technology that prevailed a year ago may no longer be adequate; that which prevailed five years ago almost certainly will not do. The success of new firms in acquiring the latest know-how is due to liberal licensing arrangements and the mobility of scientists and engineers.

Licenses and Patents

During the early fifties, the patent and licensing policies of AT&T were the only ones of importance for a firm aspiring to enter the semiconductor industry. In semiconductor technology, AT&T was the pioneer, held the strategic patents, and possessed the vital know-how.

Like other semiconductor firms, AT&T follows the general policy of patenting any new development that is patentable. Though many products and processes are obsolete by the time the patent is awarded, patents are still important. They protect the firm from inventors who later claim the development as their own. More important, a large portfolio of patents is a valuable asset in cross-licensing negotiations, making available to the firm at reduced or no royalties the important patents held by other companies.

Western Electric, which handles the licensing arrangements for AT&T, negotiates licenses on a firm-by-firm basis. With some firms, such as IBM, it has broad agreements covering many areas of electronics; with other less diversified firms, such as Fairchild, licenses are restricted mainly to semiconductors. Since an agreement normally contains a cross-licensing provision granting AT&T access to the licensee's patents, royalty payments depend on the patent portfolio and R&D capability of the licensee. Royalties charged by Western Electric varied between 0 and 5 percent of semiconductor sales until 1953, when the maximum rate was lowered to 2 percent. Usually, agreements are renegotiated after a five-year period, at which time the patent position and R&D potential of the licensee are reassessed and royalties adjusted.

Its strong patent position has not been used by AT&T to prevent firms from entering the industry. Western Electric has readily offered licenses to interested firms on reasonable terms, and Bell Laboratories has in a number of ways facilitated the early transfer of its technology to other firms. As Table 4-6 indicates, it has promptly publicized its new semiconductor breakthroughs. Its scientists and engineers have contributed many articles and books on semiconductor technology. Among the latter is William Shockley's classic book, *Electrons and Holes in Semiconductors,* published in 1950, when Shockley was with Bell Laboratories.

Bell Laboratories invites scientists and engineers from American and foreign firms to visit its facilities, observe its semiconductor operations, and discuss areas of interest with its people. During 1968, some 150 representatives of American firms toured the Bell facilities. Visits are not restricted to licensees,[23] though the latter probably benefit the most from them. Bell scientists and engineers are more likely to engage in discussions on the frontier of semiconductor technology with licensees, since any resulting discoveries are automatically available to AT&T.

Also, Bell Laboratories has conducted a number of symposia on its more important semiconductor developments. The first one on the tran-

23. Nor, on the other hand, does a license guarantee access to Bell Laboratories. Indeed, a licensee that displeases AT&T may find its plant visits curtailed.

Table 4-6. Dates of Conception, Reduction to Practice, and First Publication for Major Semiconductor Innovations Achieved by Bell Laboratories, 1947–68[a]

Innovation	*Conception*	*Reduction to practice*	*First publication*
Point contact transistor	Dec. 1947	Dec. 1947	June 1948
Zone refining	May 1950	Oct. 1950	Feb. 1952
Silicon diffusion	Feb. 1954	Feb. 1954	June 1954
Diffused base transistor (mesa type)	Dec. 1953	July 1954	June 1955
Silicon diffused base transistor	March 1955	March 1955	June 1955
Oxide masking for diffusion	June 1955	Aug. 1955	Jan. 1956
Epitaxial transistor	Sept. 1959	Feb. 1960	June 1960
Beam lead	Fall 1963	Spring 1964	Oct. 1964

Source: Correspondence with Bell Laboratories.
a. This table lists only the major innovations identified in Tables 2-1 and 2-2, none of which were introduced by Bell Laboratories after 1964.

sistor took place early in 1951 and was strictly for military and government officials. The second, attended by representatives from government agencies, universities, and more than eighty industrial firms, was held later the same year. It described transistor properties and applications, but little of the physics and technology involved. The latter were covered in an eight-day symposium held in April 1952 for Western Electric licensees. The twenty-five American firms and ten foreign firms that sent representatives each paid $25,000 in advance royalties for admission. The next symposium for licensees was in 1956, and dealt with diffusion, oxide masking, and other important developments that had occurred since 1952. After 1956, Bell Laboratories relied on individual visits by its licensees rather than symposia to pass on its new semiconductor developments.[24] Plant visits, for example, were used in late 1959 and 1960 to transfer the new epitaxial techniques.

Why has Bell Laboratories been so generous in extending technical assistance to other semiconductor firms? There are several possible reasons. The vice president for electronic components development at Bell Laboratories has stressed the following:

> There was nothing new about licensing our patents to anyone who wanted them. But it was a departure for us to tell our licensees everything we knew. We realized that if this thing [the transistor] was as big as we thought, we couldn't keep it to ourselves and we couldn't make all the technical contribu-

24. It has, however, held symposia since 1956 in other fields, such as electric switching for computers in communications equipment.

tions. It was to our interest to spread it around. If you cast your bread on the water, sometimes it comes back angel food cake.[25]

Certainly the great probability that other firms were going to use the new technology with or without licenses is another reason for the liberal licensing policy. Secrecy is difficult to maintain in the semiconductor field because of the great mobility of scientists and engineers and their desire to publish. Moreover, semiconductor firms, particularly the new, small ones, have demonstrated over and over again their disposition to infringe on patents. The prospect of lengthy and costly litigation in which its patents might be overturned could not have been very attractive to AT&T. Even if successful, such courtroom battles pitting the giant firm against small rivals damage public relations.

By offering favorable licensing arrangements, AT&T has enticed many firms, including all the major semiconductor producers, to accept its patents. As a result, it has access to the important semiconductor developments arising in other firms. And royalties, though not great in light of the major advances covered by Bell patents, are far from negligible.[26]

Finally, the antitrust suit against AT&T initiated by the Justice Department in 1949 must have influenced the company's policy of swiftly disseminating its new technology. An apparent attempt to monopolize or even dominate the new industry could easily have jeopardized its case. And the stakes were high: the government wanted Western Electric severed from the AT&T system. The 1956 consent decree ending the suit left Western Electric within AT&T, but stipulated that Western Electric license all existing patents royalty free to any interested domestic firm (though Western Electric could ask for a cross-licensing provision) and all future patents at reasonable royalties. Though a few firms did acquire Western Electric licenses on the royalty-free basis, the importance of this restriction was not great because Western Electric soon acquired a host of new patents covering diffusion, oxide masking, photolithography, epitaxy, and other important advances.

The AT&T licensing policy has prevailed throughout the semiconductor industry, largely because AT&T has acquired most of the important patents. In addition, the fear of encouraging competitors to infringe and

25. A quotation attributed to Jack Morton, appearing in "The Improbable Years," in *The Transistor: Two Decades of Progress, Electronics,* Vol. 41 (Feb. 19, 1968), p. 81.

26. Between 1952 and 1963, the firm estimates it received $9.5 million from American and foreign firms in royalties for transistor production and usage, and the amount has increased since. See "In the Matter of the Patents Act, 1949," Exhibits B, C, and D.

the Justice Department to bring an antitrust suit has kept other firms from deviating too far from AT&T's liberal licensing policy. Usually agreements are negotiated on a firm-by-firm basis and include cross-licensing clauses. A firm with few or no patents to exchange must pay royalties, anywhere from 1 to 6 percent on sales. Companies collect patents to reduce their royalty payments and increase their royalty receipts, not to restrict entry. In fact, most new firms can infringe on patents with impunity until they become important enough to make a suit worthwhile. Even then, the normal outcome is an out-of-court settlement whereby the infringing firm agrees to accept a license.

Since the early 1950s Western Electric has had the greatest number of licensees in the semiconductor industry, although for a time RCA licensees could use the essential Bell technology as a result of cross-licensing arrangements between RCA and Western Electric. A few other firms have also acquired strategic semiconductor patents that have made licensing agreements with them attractive. In the mid-1950s, Philco developed a useful family of diodes and transistors using jet-etching techniques. It offered to license other firms and was also willing to sell its automated production equipment. Several firms, including Sprague and General Instrument, did acquire Philco licenses, but the number was never large, primarily because the technology was soon surpassed by the mesa and diffusion techniques.

In the 1960s a number of firms acquired a Texas Instruments or Fairchild license. Texas Instruments holds the strategic patents on integrated circuits and Fairchild on the planar process. Although the production of integrated circuits involves technology covered by both sets of patents, since Fairchild and Texas Instruments agreed in 1966 not to dispute each other's patents, a license from one has protected a firm against a patent suit from the other. Fairchild has actively pursued companies to persuade them to take out its licenses, and charges royalties of up to 6 percent on sales. In return it provides considerable technical assistance. Texas Instruments appears less aggressive in urging firms to accept its licenses and slightly less willing to become actively involved in transferring technology to firms other than its own foreign subsidiaries.

Mobility of Scientists and Engineers

Liberal licensing policies have facilitated the entry of new firms into the semiconductor industry. They have allowed such firms to use the latest technology and even promoted its actual transfer. But alone they are not enough. New companies must have experienced scientists and engineers before the technology can be absorbed and used to compete successfully in the semiconductor market.

The interfirm mobility of scientists and engineers has always been great in the semiconductor industry. New firms have been able to tap the established firms for the necessary technical people, and also for managers, who because of the research-intensive nature of the industry are often scientists and engineers by training.

High mobility is partly reflected in the numerous instances of employees leaving established firms to set up their own. Many such spin-offs have already been noted. For further illustration, Figure 4-1 shows the firms descending from Bell Laboratories—fifteen firms and five generations in less than twenty years. While Bell Laboratories has more descendants than any other semiconductor firm, primarily because of its prolific grandchild, Fairchild, it is by no means unusual in having offspring. Hughes, Philco, RCA, Sylvania, Texas Instruments, Motorola, and others have all experienced such defections.[27]

Despite the many spin-offs, most mobility involves a move from one existing firm to another. This type of mobility facilitates the entry of firms into the industry as well as into new product lines within the industry. Former Bell Laboratories employees helped Texas Instruments and Sylvania get started in the transistor business. A number of Fairchild executives departed in 1967 to revitalize the ailing National Semiconductor Corporation. Fairchild in turn enticed C. Lester Hogan, general manager of Motorola's semiconductor division, and a number of his associates to switch companies in 1968. Many other examples could be cited, but perhaps the extent of mobility is best illustrated by the following account of the Institute of Electrical and Electronics Engineers meeting in 1966:

> The rules said "no recruiting" during the IEEE show in New York. But one company official summed up the intensity of the talent search when he said: "The show was a success. Our company lost only three men."
>
> Outside the Coliseum a shouting match took place between recruiters and IEEE officials who tried to move the recruiters away from the entrance to the show. In hotels nearby, engineers awoke to find job offers stuffed under their doors.[28]

Scientists and engineers leaving one company to join another or to es-

27. Diagrams showing the descendants of these firms are found in the source cited for Figure 4-1.

28. "Electronics Review," *Electronics*, Vol. 39 (April 4, 1966), p. 46.

Figure 4-1. Semiconductor Firms Descending from Bell Laboratories, 1952–67

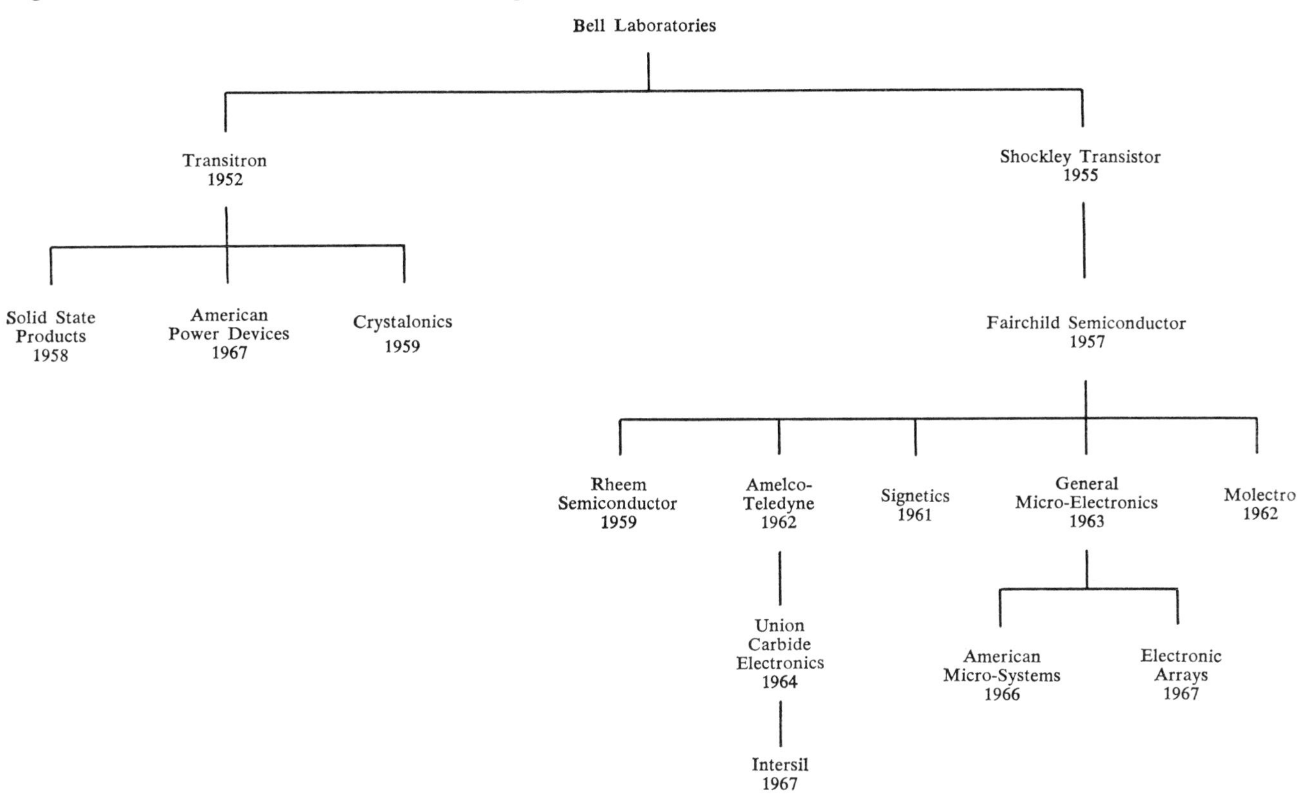

Sources: Don C. Hoefler, "Semiconductor Family Tree," *Electronic News* (July 8, 1968), pp. 4ff.; and other articles from the trade press.

tablish their own take with them not only their general knowledge about the semiconductor field but often specific expertise based on the parent company's R&D and production. Solid State Products, a spin-off from Transitron, beat its parent company to the market with a silicon controlled rectifier. When National Semiconductor was working on an improved version of a fast-selling type of integrated circuit made by Texas Instruments, it was reportedly no coincidence that the designer of that circuit left Texas Instruments and joined National Semiconductor.[29]

One major reason scientists and engineers so frequently leave good and secure jobs to join or found firms just entering the semiconductor industry is readily apparent. The two brothers who established Transitron in 1952 sold 13 percent of its stock in 1959 for over $30 million.[30] A second sale of about the same number of shares netted a similar amount a year later. The eight founders of Fairchild's semiconductor division divided over $3 million worth of Fairchild Camera and Instrument stock in 1959 when the firm exercised its option to buy the semiconductor operations it had agreed to underwrite two years earlier.[31] In addition to the potential financial rewards, one gains greater authority and more say about company operations by establishing a new firm. And seldom are the personal hardships great if the firm founders. A company employing a few competent people is often acquired by an established firm for this asset alone. All in all, the great demand for managerial and technical people throughout the industry assures many new job opportunities.

Similar incentives prompt scientists and engineers to move from one established firm to another, though the potential financial rewards are normally smaller. There are, however, exceptions. When C. Lester Hogan became Fairchild's new president in 1968, he received an annual salary of $120,000—$30,000 more than his previous salary at Motorola—and stock options that within two months of his arrival had earned $2.5 million in paper profits.[32]

29. "Electronics Review," *Electronics,* Vol. 40 (Oct. 16, 1967), p. 41.

30. "Transitron Sets Investors Agog," *Business Week* (Dec. 5, 1959), p. 123.

31. "Semiconductors," *Business Week* (March 26, 1960), p. 113.

32. "The Fight That Fairchild Won," *Business Week* (Oct. 5, 1968), p. 106. These paper profits were wiped out during the stock market decline of 1970, when Fairchild stock fell from $96 to $20. However, after canceling his stock options, Hogan was granted a $1 million interest-free loan to buy 50,000 shares at $20 each. "Recession Cools Fairchild's Recovery," *Business Week* (Jan. 30, 1971), p. 84.

While the incentives encouraging scientists and engineers to change jobs are easy to discern, why the established firms have been so tolerant of this movement is more difficult to comprehend. Fairchild did sue its former employees who established Rheem Semiconductor, charging they took with them proprietary material.[33] And Motorola brought action against Hogan and its other executives who joined Fairchild in 1968.[34] But such steps are rare.

Here, as with patents and licenses, Bell Laboratories set the precedent for the industry's behavior. During the 1950s, it emerged as the major source of semiconductor scientists and engineers, and has since served as the principal training ground for numerous semiconductor specialists. Its former employees, commonly called Bell graduates, are found in nearly all the important semiconductor firms.

For several reasons, Bell Laboratories maintains a lenient attitude toward personnel defections. First, many Bell employees receive lucrative offers which Bell Laboratories usually cannot match since the growth of AT&T's semiconductor operations is restricted by the 1956 consent decree. An attempt to keep them anyway would create ill will, damage personnel morale, and reduce the attractiveness to prospective employees of working at Bell Laboratories. Second, it is very difficult, if not impossible, to discourage mobility through legal actions alleging the illegal use of proprietary material. Third, Bell Laboratories receives useful information through the friendly, informal relations it maintains with former employees scattered throughout the industry. This feedback is no doubt facilitated by the fact that Western Electric has cross-licensing agreements with many semiconductor firms and does not compete in the commercial market. Finally, a lenient attitude toward personnel mobility is consistent with AT&T's liberal licensing policy.

Bell's attitude has permeated the industry. Though a few firms, particularly Texas Instruments and IBM, strenuously try to minimize defections, their approach is to make working conditions and opportunities so attractive that few are motivated to leave. Rarely are departing employees threatened with legal action.

33. "Semiconductors," Special Report, *Business Week* (March 26, 1960), p. 113.

34. "The Fight That Fairchild Won," p. 106; "High Noon," *Forbes* (Feb. 15, 1971), p. 26.

Economies of Scale, Learning Economies, and Capital Requirements

The relatively modest amount of capital needed to enter the semiconductor industry is another major factor facilitating entry. Since capital requirements depend on economies of scale and learning economies, this section first examines their importance in the semiconductor industry. It then assesses the amount of capital needed for successful entry and the availability of venture capital.

Economies of Scale

Economies of scale permit average costs to fall as output rises.[35] They may arise in R&D, production, or marketing activities. In addition, the vertical integration of semiconductor operations with the production of final electronic products may generate such economies.

The R&D effort needed to maintain proficiency in all aspects of semiconductor technology has increased in size and cost as the frontiers of semiconductor technology have pushed out and expanded.[36] One manager estimates that a company must now spend $10 million to $15 million a year just to stay abreast of the latest technology.[37] Since few firms can long afford to spend for R&D more than 10 percent of sales, only the very largest producers can support such an effort.

Yet for several reasons many new firms have successfully entered the industry with very little initial R&D. Many have used borrowed technology, particularly the technology based on the R&D efforts of parent

35. More precisely, economies of scale permit average costs to fall as output for a given production period increases. Cost reductions that arise as cumulative output increases are generally associated with learning economies.

36. Elsewhere we have argued that this is the typical pattern followed by new industries until their technology stabilizes and production techniques become standardized. See Dennis C. Mueller and John E. Tilton, "Research and Development Costs as a Barrier to Entry," *Canadian Journal of Economics,* Vol. 2 (November 1969), pp. 570–79.

37. C. Lester Hogan reportedly made this statement while still general manager of Motorola's semiconductor division. He may have exaggerated the figure to scare prospective new firms. See "On the Right Track," *Electronics,* Vol. 41 (May 13, 1968), p. 39.

firms. And across-the-board proficiency in semiconductor technology has not been necessary. Often the capability for producing one good semiconductor device, usually a new one, is sufficient.

Once established in one product line, a new firm can support a larger R&D program. At this point, many opt to remain specialty houses and concentrate their R&D efforts on a few selected products. Others, such as Texas Instruments and Fairchild, choose to diversify into a wide range of products. This strategy naturally requires a more extensive and expensive R&D program, though it can grow at a pace that expanding sales can support.

In production, large economies of scale have not yet materialized. Most firms have hesitated to install expensive automated equipment, which the swift pace of technological change may make obsolete. The typical product life cycle is amazingly short. Over half the transistor types introduced during the 1956–58 period were on the market less than two years. Though the obsolescence rate declined somewhat in the 1960s, it is still very high compared to other industries.[38] In these circumstances, firms have preferred to rely on relatively labor-intensive production techniques. As the technology stabilizes, automated equipment may be more widely introduced, bringing about an increase in economies of scale. Even so, the industry is not likely to become highly capital intensive.

One firm that did automate was Philco. After innovating the jet-etching process in the mid-fifties and producing with it the surface-barrier transistor and other related semiconductor devices, it developed its Fast Automatic Transfer Line, the first really automated production facility in the industry. This equipment reduced costs, allowing Philco to slash prices and capture over 70 percent of the high-frequency alloy-transistor market. But the advent of mesa, diffusion, and planar techniques made Philco's technology and equipment obsolete within a few years. Meanwhile the company had become attached to its own family of devices and production techniques. Reluctance to abandon the old technology for the new soon led to its downfall.

In marketing, the opportunities for scale economies are few. Potential semiconductor customers in the American market numbered only about

38. Jerry Eimbinder, "Transistor Industry Growth Patterns," *Solid State Design*, Vol. 4 (January 1963), Fig. 7; "Where Time Moves at a Dizzying Pace," *Business Week* (April 20, 1968), p. 174.

4,000 in 1968, a few hundred of which accounted for 95 percent of all sales,[39] so that finding the customers is relatively easy. Most buyers are final electronic equipment producers who are not susceptible to massive advertising campaigns. Nor does the nature of the product require a large servicing network.

Some costs are incurred in acquainting circuit designers with the potential uses and properties of new semiconductor devices. Texas Instruments built the first radio using transistors and the first computer using integrated circuits to demonstrate their feasibility. Recently, as certain series of integrated circuits developed by Fairchild and Texas Instruments have become widely accepted by circuit designers, other firms have had to undertake promotional efforts to get their own integrated circuits accepted. Though this may inhibit entry in the future, it has not been a major barrier in the past.

The vertical integration of semiconductor and final equipment production is another possible source of scale economies. Although many semiconductor firms do produce final electronic products, they apparently do not derive an important competitive advantage over other firms as a result. Indeed, too close a tie between the two activities can adversely affect the competitive position of both. Product divisions are hurt if pressured to use in-house semiconductors when for their purposes better or cheaper ones are available from outside firms, and the semiconductor division suffers if forced to expend more development effort on devices desired by sister divisions than justified by its overall market strategy. As a result, most integrated firms give their semiconductor and product divisions considerable autonomy, allowing the latter to buy components outside the firm and the former to turn down an order placed by a sister division. This often produces the bizarre situation where product divisions of two companies are competing vigorously against each other in the market while one is buying components from the other's semiconductor division.[40]

The integrated circuit, since it contains a complete circuit in one device, encroaches to some extent on the final equipment producer's domain and reduces the value added by the latter to the final system. It

39. Estimate of the vice president for semiconductor marketing at Texas Instruments; see "Where Time Moves at a Dizzying Pace," p. 178.

40. P. E. Haggerty, "Integrated Electronics and Change in the Electronics Industry" (speech delivered to the International Electron Devices meeting, Oct. 19, 1967; processed), p. 14.

also increases the degree of coordination necessary between the producers of components and of systems. The systems producer's reliance on the semiconductor producer is thereby increased and its ability to protect proprietary designs reduced. Though these considerations provide a strong incentive for systems firms to set up their own integrated-circuit facilities, most have continued to rely heavily on the several major integrated-circuit companies, using in-house capability primarily to build prototype circuits and provide integrated-circuit expertise to advise management. So even in integrated-circuit production the benefits from vertical integration do not appear to offset the disadvantages. Among the latter, the failure of nearly all captive semiconductor operations to realize learning economies to the same extent as the independent producers is particularly important.

Learning Economies

Unlike economies of scale, learning economies are important in the semiconductor industry. From past experience, firms have determined that the average cost of producing a semiconductor device normally falls between 20 and 30 percent every time its cumulative past production doubles.[41]

Several characteristics of these learning economies should be noted.[42] First, learning by doing on the part of production workers is only one, relatively unimportant, source of such economies. Most are generated by improvements in the production process arising from experimentation with times and temperatures used in the preparation of semiconductor crystals, the measures taken to control crystal contamination during fabrication, the equipment used in assembling and testing, the scheduling

41. These figures were obtained in interviews with semiconductor firms. In addition, see the Boston Consulting Group, *Perspectives on Experience* (Boston: BCG, 1970), particularly Exhibits 14–18; and "Texas Instruments: 'All Systems Go,'" *Dun's Review*, Vol. 89 (January 1967), pp. 28ff.

42. These characteristics apparently are not unique to learning in the semiconductor industry. See Richard R. Nelson and Sidney G. Winter, "Production Theory, Learning Processes, and Dynamic Competition" (unpublished paper, no date).

Many studies take a narrower view of the learning process than that taken here or by Nelson and Winter. Examples of such studies along with an argument favoring a wider interpretation of the learning process are found in Nicholas Baloff, "The Learning Curve—Some Controversial Issues," *Journal of Industrial Economics*, Vol. 14 (July 1966), pp. 275–82.

and organization of operations, and so on. Second, most learning economies are not an automatic by-product of production, but rather are produced by the deliberate efforts of firms. Engineers and other specialists are assigned to study the production systems of new devices and to experiment with possible improvements. A considerable portion of R&D activity is devoted to improving the manufacturing process and the characteristics of semiconductors already in production. Third, learning economies, though identified with cumulative past production by the industry, may also be a function of time in production. Certainly, some of the learning activities just described depend as much on time as on production activity to generate improvements.[43] Fourth, although some learning readily becomes general knowledge and thus a public good, much is either uniquely applicable to a particular operation or can be transferred to another facility only with technical assistance from the firm having the know-how. Consequently, a large portion of the benefits produced by learning accrues to the firms doing the learning.

The importance of learning is recognized throughout the industry and taken into account by firms in bidding on large semiconductor orders. This occasionally leads to the curious situation where a company agrees to supply a device at a price below current costs and still makes money on the order. Motorola provides an example. About 1960 it contracted with Chrysler Motors to supply silicon rectifiers for automobile alternators for seventy-five cents apiece when the prevailing industry price was around two dollars. Early runs did not cover costs, but production techniques so improved while the large order was being filled that Motorola managed to make money on the contract. The firm has remained a strong contender in this market ever since.[44]

Substantial learning economies bestow on early producers of new semiconductor devices an appreciable cost advantage over later entrants. However, at least in the past, they have not posed a great barrier to entry. The swift pace of technological change has repeatedly forced firms to abandon old devices—some only a year or two old—and to undertake the

43. In practice, the effects of time in production and cumulative output on learning are difficult to separate. For one interesting attempt suggesting that time is more important than cumulative output, see William Fellner, "Specific Interpretations of Learning by Doing," *Journal of Economic Theory,* Vol. 1 (August 1969), pp. 119–40.

44. See Ed Patterson, *DE: Semiconductors* (London: *Design Electronics,* 1967), p. 14.

production of new ones. At such times, the learning economies enjoyed by the established firms are lost, and new firms can generally enter without great handicaps. In fact, their lack of experience and vested interests in the obsolete technology may be an asset.

Most new companies, and particularly the successful ones, have initially concentrated on new product lines: Transitron on gold-bonded diodes, Texas Instruments on silicon transistors, Fairchild on mesa and planar transistors, General Micro-Electronics and General Instrument on MOS transistors, and Signetics on integrated circuits. New firms have emphasized new products because they suffer no competitive disadvantages from learning economies in these areas. Furthermore, new products often offer large potential profits—Schumpeterian entrepreneurial profits based on an early technological lead—that can be used to expand the company or to entice a large firm to buy it at a handsome profit to the founders.

Another reason learning economies have not impeded entry arises from the fact that many new semiconductor firms are spin-offs from established firms. To some extent employees carry with them the production know-how of the parent firm. Just how much accumulated learning can be transferred from one firm to another is difficult to assess. The success of American subsidiaries in penetrating the European market, described in the next chapter, suggests that a fair amount can be transferred. Still, few new companies try to compete with their parent firm in product lines in which the latter is well established and has substantial past production. Rather they prefer to enter areas where the parent is planning to enter or has just entered. In such cases, the spin-off can borrow from the parent's R&D efforts, but not from the know-how generated by learning economies since the parent's own production experience is small or zero.

Entry Costs and the Availability of Venture Capital

So far in the young semiconductor industry's history, economies of scale have not been great, and learning economies, though appreciable, have continually been wiped out by technological change. As a result, new firms have successfully entered the industry without huge outlays of capital. General Transistor, which for awhile in the early 1950s ranked second only to Raytheon in transistor production, got into the business with a $100,000 investment, and $1 million launched Transitron.[45] We have already noted that Texas Instruments committed just over $4 million before its semiconductor operations made a profit. Fairchild Camera and Instrument reportedly did not suffer unusual expenses of more than $350,000 a year during 1957–58, the two years it financed the spin-off group from Shockley Laboratories before its semiconductor activities began making money.[46]

Those familiar with the industry generally agree that the cost of entry rose in the 1960s, though they differ on the actual amounts required.[47] While increasing capital requirements may already be slowing the pace of entry and may be an even more serious barrier in the future, they are not yet beyond the reach of many small firms, as the emergence of Signetics, General Micro-Electronics, Intersil, and others testifies. In fact, capital apparently is more of a problem for the successful small company that wants to expand rapidly. Signetics and General Micro-Electronics, for example, merged with larger firms after they were established, and in Signetics' case at least, capital was an important consideration.[48]

While entry has been possible with relatively small capital outlays for many companies, the admission price for all companies has not been low. Many of the receiving tube firms have invested much more than the amounts just noted, and yet achieved only moderate success in the market. Motorola committed a substantial part of the $22 million it spent on capital investment in 1963 to its integrated circuit operations, and did not anticipate any profit on this investment before 1965.[49]

Firms entering the semiconductor industry have not, at least until re-

45. William B. Harris, "The Battle of the Components," *Fortune* (May 1957), p. 138; "Semiconductors," *Business Week* (March 26, 1960), p. 110.

46. Ibid., p. 113.

47. Estimates ranging from $1.5 million to $20.0 million, for instance, have been cited as necessary to enter the integrated-circuit business successfully. See Arthur D. Little, Inc., *Trends in Integrated Circuits and Microelectronics* (Cambridge, Mass.: ADL, 1965), p. 28; Barry Miller, "Competition to Tighten in Semiconductors," *Aviation Week and Space Technology* (Sept. 23, 1963), p. 67; and Walter E. Weyler and others, *Integrated Circuits: Technical Review and Business Analysis* (Cambridge, Mass.: Integrated Circuit Associates, 1963), pp. 80–81.

48. Miller, "Competition to Tighten in Semiconductors," p. 61; "Intersil: Upstart with Talent," *Business Week* (Sept. 12, 1970), pp. 74ff.

49. James Lydon, "No Integrated Circuits Profit Seen at Motorola Before '65," *Electronic News* (Dec. 16, 1963), p. 31.

cently, encountered difficulties in finding venture capital. Transitron's initial capital was provided by one of its founders. Texas Instruments diverted funds from other activities to start its semiconductor operations and then plowed back semiconductor profits for expansion. The eight scientists from Shockley Laboratories were backed by Fairchild Camera and Instrument, which was looking for a new area in which to diversify following cutbacks in government contracts in the photographic field. Signetics' initial financing came from an investment banking house, and capital for expansion was provided by Corning Glass.

Venture capital for new semiconductor firms has been readily available, at times even lavishly so. This may now be changing. The debacle of the early sixties, when semiconductor prices and profits plummeted, coupled with the growing amounts of capital necessary has made investors more wary. Nevertheless, so far venture capital has posed no great barrier to entry.

The Role of Government

The government has influenced the entry of new firms and the development of the semiconductor industry in a host of ways—the antitrust activities of the Justice Department, the decision by the military not to classify the transistor, and many more. This section examines two government activities that have greatly affected the industry—the military demand for semiconductors and the financial support the government has given firms for R&D and production facilities.

Military Demand

The importance of the defense market for semiconductors is shown in Table 4-7. This market grew from $15 million in 1955 to $294 million in 1968 and, depending on the year, accounted for between one-fourth and one-half of the total market.

The impact of military demand on the semiconductor industry transcends its size. The armed forces have always imposed the most rigid standards and quality control. They have constantly demanded better devices and have not hesitated to inform the industry of specific needs. Moreover, they provide a substantial market for new devices that meet their requirements.

Table 4-7. U.S. Production of Semiconductors for Defense Requirements, 1955–68[a]

Year	*Total semiconductor production (millions of dollars)*	*Defense semiconductor production[b] (millions of dollars)*	*Defense as a percentage of total*
1955	40	15	*38*
1956	90	32	*36*
1957	151	54	*36*
1958	210	81	*39*
1959	396	180	*45*
1960	542	258	*48*
1961	565	222	*39*
1962	575	223	*39*
1963	610	211	*35*
1964	676	192	*28*
1965	884	247	*28*
1966	1,123	298	*27*
1967	1,107	303	*27*
1968	1,159	294	*25*

Sources: Data for discrete devices are from U.S. Department of Commerce, Business and Defense Services Administration (BDSA), *Electronic Components: Production and Related Data, 1952–1959* (1960); BDSA, "Consolidated Tabulation: Shipments of Selected Electronic Components" (annual reports; processed; title varied somewhat over the period).

a. The 1962–68 data include monolithic integrated circuits. Figures on the latter are as shown in Table 4-8 and come from the sources given there.

b. Defense production includes devices produced for Department of Defense (DOD), Atomic Energy Commission (AEC), Central Intelligence Agency (CIA), Federal Aviation Agency (FAA), and National Aeronautics and Space Administration (NASA) equipment.

The latter is particularly important. Often new and better semiconductors are initially too expensive for industrial or consumer electronic products. In military equipment, reliability and performance have priority over costs, so that most new semiconductor devices first find a home in military products. As production proceeds, learning occurs and costs fall. Within a few years, the price is low enough to penetrate the industrial market, and eventually the consumer market.

This typical shifting market pattern is illustrated for integrated circuits in Table 4-8. In 1962, the year after they were introduced, their average price was about $50.00, too high for use in commercial products. By 1968, the price had fallen to $2.33 and the military's share of total output had dropped to 37 percent.[50] Integrated circuits were widely used in computers and other industrial products and were being considered for

50. Some of the price decrease, of course, was caused by (rather than being the cause of) the increase in relative importance of the commercial market, which does not demand as high performance standards as the military market.

Table 4-8. U.S. Integrated-Circuit Production and Prices, and the Importance of the Defense Market, 1962–68

Year	*Total production (millions of dollars)*	*Average price per integrated circuit (dollars)*	*Defense production as a percentage of total production*[a]
1962	4[b]	50.00[b]	*100*[b]
1963	16	31.60	*94*[b]
1964	41	18.50	*85*[b]
1965	79	8.33	*72*
1966	148	5.05	*53*
1967	228	3.32	*43*
1968	312	2.33	*37*

Sources: Total production and average price figures are from the *Electronic Industries Yearbook, 1969* (Washington: Electronic Industries Association, 1969), Table 55. Defense production as a percentage of total production is based on data for monolithic integrated circuits found in BDSA, "Consolidated Tabulation: Shipments of Selected Electronic Components."

a. Defense production includes devices produced for DOD, AEC, CIA, FAA, and NASA equipment.

b. Estimated.

radios and consumer products.[51] The few exceptions to this typical pattern involve primarily new semiconductors whose virtue is lower costs rather than improved reliability or greater capabilities. One example is plastic-encapsulated devices, which the armed forces have hesitated to accept for fear they will prove less reliable than devices using the conventional and more expensive seals. Generally, however, it is the military that first uses new semiconductors and provides the immediate incentive for firms to develop them.

The defense market has been particularly important for new firms. For reasons noted in the previous section, these firms often have started by introducing new products and concentrating in new semiconductor fields where the military has usually provided the major or only market. Fortunately for them, the armed forces have not hesitated to buy from new and untried firms. In early 1953, for example, before Transitron had made any significant sales, the military authorized the use of its gold-bonded diode. This approval has been called the real turning point for the new firm.[52] During 1959, new firms accounted for 63 percent of all semiconductor sales and 69 percent of military sales.[53]

51. For a brief description of how this shifting market pattern evolved for the surface-barrier transistor, see David Allison, "The Civilian Technology Lag," *International Science and Technology* (December 1963), p. 30. For more general information on the shifting market pattern in the semiconductor industry, see OECD, *Electronic Components: Gaps in Technology,* pp. 62–65.

52. Harris, "The Company That Started with a Gold Whisker," p. 140.

53. See Table 4-10 and source.

Military demand has therefore stimulated the formation of new companies and encouraged them to develop new semiconductors by promising the successful ones a large market at high prices and good profits. Further, the military market, by activating learning economies, often serves as a stepping stone to eventual penetration into the commercial market.

R&D and Production Support

From the early 1950s on, the potential military significance of the transistor was appreciated in the government. By 1953, the Army Signal Corps was financing pilot production lines for transistors and related devices at five sites operated by Western Electric, General Electric, Raytheon, RCA, and Sylvania.[54]

Since then the government has funded the semiconductor activities of companies in three principal ways. First, it has contracted directly for R&D projects. Second, it has financed production refinement programs for transistors as well as diodes and rectifiers through industrial preparedness studies. Finally, some of its appropriations for new weapon systems are passed on by prime contractors to semiconductor firms and used for R&D or production improvements.

Direct government funding for R&D and production refinement projects is shown in Table 4-9 for 1955–61. Support for R&D accounted for more than half the funds for the period as a whole and increased substantially. A large part of the money for refinement projects came in 1956 when $14 million was appropriated for transistors. Contracts placed with about a dozen firms called for the delivery of some thirty different types of germanium and silicon transistors over the following several years. In addition, for each type, production lines capable of turning out 3,000 units a month were to be developed. While the companies paid for plant facilities, the government covered engineering design and development. This support helped firms get into the industry and greatly expanded semiconductor production capacity in the United States. Funds for diodes and rectifiers, though more modest, also contributed to this capacity.

The Department of Defense conducted a special survey in 1960 to determine the total amount of semiconductor R&D financed directly as well as indirectly by the government through its expenditures on weapon sys-

54. Francis Bello, "The Year of the Transistor," *Fortune* (March 1953), pp. 128ff.

Table 4-9. U.S. Government Funds Allocated Directly to Firms for Semiconductor Research and Development and for Production Refinement Projects, 1955–61

Use of funds	*1955*	*1956*	*1957*	*1958*	*1959*	*1960*	*1961*
Research and development	3.2	4.1	3.8	4.0	6.3	6.8	11.0
Production refinement							
Transistors	2.7	14.0	0.0	1.9	1.0	0.0	1.7
Diodes and rectifiers	2.2	0.8	0.5	0.2	0.0	1.1	0.8
Total	8.1	18.9	4.3	6.1	7.3	7.9	13.5

Source: BDSA, *Semiconductors: U.S. Production and Trade* (1961), Table 8.

tems. It found $13.9 million had been spent in 1958 and $16.2 million in 1959[55]—according to Table 4-9, more than double the amount the government funded directly through R&D contracts.

The survey also revealed that government-sponsored R&D has constituted a large part of the total in the semiconductor industry—25 percent in 1958 and 23 percent in 1959.[56] Such support is almost entirely for R&D on military requirements and directly helps recipient firms expand their military sales—an important consideration in an industry where the military accounts for between 25 and 50 percent of the total market.

Government R&D funds also help companies in the commercial market. Indeed, there is some evidence suggesting that the spillover may be substantial (see Appendix). Although conflicting with much of the prevailing opinion about the importance of spillover in general, the existence of considerable spillover in the semiconductor industry is not surprising, since similar technology is used to produce both military and commercial devices and modified versions of many devices initially used in military equipment are eventually used in industrial and consumer equipment too.[57]

Since government funds for R&D and production projects have in general helped firms, ascertaining whether the distribution of government funds has favored the large established firms or the small new firms is

55. BDSA, *Semiconductors: U.S. Production and Trade,* Table 9.

56. Ibid.

57. Although they maintain that spillover in general is not important, Nelson, Peck, and Kalachek do point out that it may be significant in certain fields and specifically mention components as a possibility. See Richard R. Nelson, Merton J. Peck, and Edward D. Kalachek, *Technology, Economic Growth, and Public Policy* (Brookings Institution, 1967), pp. 82–85.

Table 4-10. Distribution of Government and Company Research and Development Funds, and Semiconductor Sales in the United States, by Type of Firm, 1959

	Government R&D funds		*Company R&D funds*		*Semiconductor sales*	
Type of firm	*Millions of dollars*	*Percent*	*Millions of dollars*	*Percent*	*Millions of dollars*	*Percent*
Western Electric and 8 receiving tube producers	12.7	*78*	27.2	*50*	149.5	*37*
New firms	3.5	*22*	26.8	*50*	252.1	*63*
Total	16.2	*100*	54.0	*100*	401.6	*100*

Source: U.S. Department of Defense, Survey of 64 Semiconductor Companies, 1960, unpublished tabulations.

important. Unfortunately, data on production-refinement support by recipient are not available, so no conclusion about that type of assistance is possible. There are data for R&D, though again only for 1959. They indicate that the size of government R&D support increases with a firm's semiconductor sales (see Appendix). Thus the larger semiconductor producers receive more government funds, which in turn helps them maintain their market leadership.[58]

In addition to the larger semiconductor firms, the government also appears to favor the older producers. Western Electric and the receiving tube producers, early entrants in the transistor business, received 78 percent of the government funds spent in 1959 for semiconductor R&D, while accounting for only 37 percent of semiconductor sales (see Table 4-10). Using sales in this case to judge R&D capability may be misleading. We argued earlier that the receiving tube firms have made a greater contribution to the innovation process relative to the diffusion process than have the new firms. And certainly sales underestimate the R&D capability of Bell Laboratories, since Western Electric is prohibited from selling in the commercial market. As a result, Table 4-10 also indicates the amount of their own funds these firms spent on semiconductor R&D

58. Not all government R&D contracts, however, have been a boon to the firms that have won them. For instance, RCA began selling integrated circuits about 1964, several years later than its major competitors. One reason for its slow start was that the Army Signal Corps had chosen RCA as its prime contractor to develop micromodules, which proved to be an inferior technology compared with integrated circuits.

as an alternative measure of R&D capacity. Together AT&T and the receiving tube firms accounted for half of the company funds used for semiconductor R&D. This is still substantially less than their share of government R&D funds, indicating that the distribution of government R&D funds has favored the older semiconductor firms.

Unlike the military demand for semiconductors, which has stimulated the formation of new companies, government R&D funding programs, by favoring the larger, older firms, have impeded entry; however, impact on entry has apparently not been great, since other factors have more than compensated. This raises an important policy question: How important has government R&D funding been compared with government procurement in stimulating the growth and technological development of the semiconductor industry in the United States? Though evidence is limited, government procurement has probably had a much greater effect. As the tables of this section indicate, the defense market has pumped many more dollars into the industry than the government has in supporting R&D and production facilities. Moreover, many of the major semiconductor innovations were achieved by firms without any government assistance. Bell Laboratories produced the first transistor with its own funds. And despite the many millions of R&D dollars the Air Force spent to develop integrated circuits, company-financed R&D projects produced the major breakthroughs.[59] Only after Texas Instruments achieved a working model of an integrated circuit did it receive an Air Force contract for subsequent development. And Fairchild developed the planar process, which led to the mass production of integrated circuits, without any government support.

Conclusions

Since Bell Laboratories introduced the transistor in the early 1950s, many firms have entered the semiconductor industry and contributed to its development. These firms can be separated into three groups: (a) AT&T, including Bell Laboratories and Western Electric, which generated much of the industry's technology even though prohibited from selling in the commercial market; (b) the receiving tube firms, which responded to the transistor challenge by moving quickly into its production; and (c)

59. See Kleiman, "The Integrated Circuit," pp. 172–215; OECD, *Electronic Components: Gaps in Technology*, Chap. 4.

the new firms, which were new to the active-component field and so had no vested interest to protect. The latter were also new in the sense that most entered semiconductor production after Western Electric and the receiving tube firms. The new firms have become the most numerous; because of low entry barriers, they are continually entering the industry, though many survive only a few years.

All three types of firms have contributed to the rapid development of the American semiconductor industry, though in somewhat different and generally complementary ways. Bell Laboratories and the receiving tube firms have greatly promoted the industry's technology, and Bell has trained many of the scientists and engineers now working throughout the industry. Its liberal attitude toward licensing and the mobility of semiconductor professionals have served as a precedent for the entire industry. The new firms, though they too have added to the technology, have made their greatest contribution in diffusing new semiconductor products and processes. Adept at taking new developments from the laboratory—often some other firm's laboratory—and exploiting them in large-scale production, these firms have won the top sales positions in the industry and in the process assured the rapid utilization of new advances in semiconductor technology. Moreover, whenever the established firms have faltered, other new firms have come forward with important new devices, forcing the former to catch up quickly or yield part of their market to the newcomers.

Access to the new technology emanating from other firms made it possible for new firms to enter the industry and promote rapid diffusion. The liberal licensing policy prevalent in the industry not only allows new firms to use patented technology, but even facilitates its transfer. And scientists and engineers, who continually leave one firm to join another or to start their own, take with them the skills, knowledge, and know-how acquired at their former company.

Relatively low capital requirements along with ample venture capital have also helped new firms move into the semiconductor industry. Economies of scale have not been great, primarily because the technology has changed swiftly. Firms hesitate to invest heavily in automated equipment that may soon be obsolete.

The rapid pace of technology has also prevented the substantial learning economies that characterize production in this industry from becoming a major barrier to entry. By concentrating on new products, where the production experience of all firms is negligible, new firms have entered the industry without suffering a competitive disadvantage from the benefits that learning economies bestow on the established firms.

Government funds for R&D and production facilities have accelerated the development of the American semiconductor industry but hindered entry by helping the larger, older semiconductor firms maintain and expand sales in military and commercial markets. The large defense market for semiconductors, which accounts for a significant segment of the overall market, appears to have played a more important role in fostering the industry's growth and technological development. It demands the best quality semiconductors. It offers the financial incentives that stimulate the development of better devices for defense equipment, and eventually for commercial products as well, for in filling military orders learning economies arise and reduce costs. The defense market is particularly important for new firms hoping to introduce new semiconductor devices and has facilitated their entry.

CHAPTER FIVE

Europe

BRITAIN, FRANCE, AND GERMANY produce most of Europe's semiconductors. This chapter examines the firms responsible for introducing and diffusing new semiconductor technology in these countries. It begins by identifying the different types of semiconductor firms and then proceeds to analyze their contributions to the innovative and diffusion processes. The initial sections reveal that in many respects the European firms producing receiving tubes have responded to the transistor and the new semiconductor technology in about the same way as their American counterparts. The important functions performed by the new firms in the United States, however, have been largely assumed in Europe by American subsidiaries rather than new European firms. The latter are much less common and with few exceptions have remained small and specialized. Why the barriers to entry are high for European but not for foreign firms is the principal concern of the last three sections, which examine licensing policies, mobility of scientists and engineers, availability of venture capital, market size, economies of scale, learning economies, military demand, government research and development (R&D) support, and industry rationalization programs.

Types of Firms

Semiconductor firms active in the European market can be separated into three groups—receiving tube producers, new foreign subsidiaries, and new firms.[1]

1. Since imports account for a larger share of the market in Europe than in the United States, one might also add foreign firms without subsidiaries which export to the European market. However, the major exporters to Europe have established manufacturing subsidiaries there. In any event, firms do not contribute to the diffusion of semiconductor technology on the producer level through exports.

Receiving Tube Producers

The only significant receiving tube producers in England in the post-war period have been Mullard, Associated Electrical Industries (AEI), and Standard Telephones and Cables (STC). Mullard, which dominates the market, is a subsidiary of Philips, a huge multinational firm whose home office is in the Netherlands. Standard Telephones and Cables is a subsidiary of International Telephone and Telegraph (ITT), an American firm, though most of its electronic activities are outside the United States. Associated Electrical Industries is controlled by British interests. In 1967, the firm was acquired by the British General Electric Company (GEC), which is independent of the American firm with the same name. Then a year later, GEC merged with English Electric (which in 1967 had acquired Elliott-Automation) to form an all-British electrical firm that ranks among the largest in the world.

In France, the receiving tube firms include Radiotechnique Compelec (RTC), Compagnie Générale d'Electricité (CGE), and Thomson-CSF. Radiotechnique Compelec, like Mullard, is a Philips subsidiary; the other two are controlled by French interests. Thomson-CSF was formed in 1967 by the merger of Compagnie Française Thomson-Houston (CFTH) and Compagnie Générale de Télégraphie Sans Fil (CSF), both of which produced receiving tubes.

In Germany, Siemens, AEG-Telefunken, Standard Elektrik Lorenz, and Valvo are the receiving tube producers. Valvo is the German subsidiary of Philips, and Standard Elektrik Lorenz of ITT. Telefunken is part of the Allgemeine Elektrizitäts Gesellschaft (AEG) group, which, like Siemens, is controlled by German interests.

Receiving tube producers in Europe as in America are divisions or subsidiary companies of large diversified firms. Philips, ITT, Siemens, English Electric–GEC, Thomson-CSF, CGE, and AEG-Telefunken—all produce a wide range of components as well as final products and derive only a fraction of their overall revenues from receiving tube sales.[2] As a result, the transistor and other semiconductor devices have not posed a serious threat to their survival or growth.

2. Of course, for some of the subsidiary companies—Mullard, for example—receiving tube production constitutes a major part of their overall operations.

New Foreign Subsidiaries

This group of firms contains the manufacturing subsidiaries *established in the postwar period* by foreign semiconductor firms. Two kinds of subsidiaries are not included in this group. The first covers the affiliates of Philips and ITT that are counted among the receiving tube firms. These subsidiaries existed before World War II and the advent of the transistor. They are managed by nationals, enjoy considerable autonomy, and behave more like established native firms than new foreign subsidiaries. Indeed, the newcomers create just as great a problem for them as for the native firms. For these reasons, the older subsidiaries are accepted as part of the established industrial order and viewed with far less concern.

Firms acquired by foreign companies after entering the European semiconductor industry constitute the second kind of subsidiary excluded from this group. One example is Intermetall, which Clevite bought in 1955 and then sold ten years later to ITT. Another is Compagnie des Dispositifs Semiconducteurs Westinghouse (CDSW), in which the American firm Westinghouse has an 82 percent interest and which was formed in 1967 when Westinghouse purchased the rectifier division of Compagnie des Freins et Signaux Westinghouse (which is independent of Westinghouse), and combined it with the silicon department of Société Jeumont-Schneider, co-owner of the new firm. Acquired subsidiaries like Intermetall and CDSW are not common in the European semiconductor industry, and the few that do exist behave more like the new European firms (which they were originally) than the new foreign subsidiaries. Consequently, they are included with the former. (The one exception is Intermetall, which after 1965, the year Clevite sold it to ITT, is included with the receiving tube firms.)

Texas Instruments, IBM, Società Generale Semiconduttori (SGS), and Motorola have the most important new foreign subsidiaries in Europe. Transitron, Sprague, Microwave Associates, International Rectifier, and Hughes have also established semiconductor production facilities there, and Signetics, National Semiconductor, General Instrument, and Union Carbide are in the process of doing so. Most of these companies are successful new firms in the United States and are trying to capitalize on their technology and know-how in additional markets. Only a few deviate in one respect or another from this typical genre.

Though an Italian company, SGS is not the exception it may at first appear to be. Between 1961 and 1968, the period when SGS established its British, French, and German subsidiaries, Fairchild held a one-third interest in the firm. The American company provided crucial production expertise and marketing skills, and helped SGS establish its own R&D facility. Problems in its American operations and growing differences with its Italian partners prompted Fairchild to sell out in 1968.[3]

In another way, IBM also differs from the other firms. Its European semiconductor facility located in France produces only for in-house use and therefore does not compete with other semiconductor firms in the market. Indeed, IBM is a major purchaser of integrated circuits and discrete devices, needed for the hybrid circuits used in its 360 series of computers. For this reason, as well as the lack of data concerning its semiconductor output, IBM is often not counted as part of the French semiconductor industry.

Another unusual firm is Microwave Semiconductor Devices, a subsidiary of Microwave Associates that produces devices for microwave applications in the American market. The American firm established its British subsidiary in 1961, not to exploit in Europe technology used successfully in the United States, but instead to expand its operations into the high-volume general semiconductor market. It had determined that the technical personnel needed for this expansion program could be obtained at less cost in England. The electronic recession in the United States plus competition for the British planar-transistor market from SGS-Fairchild and STC caused the firm to abandon the attempt a year later.

New European Firms

This category contains the other firms that have manufactured semiconductors in Britain, France, and Germany. These firms can be separated into subgroups similar to those identified for new firms in the American industry. First, there are companies like English Electric that were large and diversified when they entered the semiconductor industry. Most of these firms are in England. Some had once produced receiving tubes but been pushed out of market; semiconductors offered another opportunity in the active-components field. Second, there are the medium and small firms established in other, often related, fields. An example is

3. At that time, Olivetti, Telettra, and Fairchild each owned one-third interest in SGS. Telettra also sold its interest to Olivetti shortly after Fairchild did.

Table 5-1. Semiconductor Firms in Great Britain and Years They Were Active in the Industry, 1954–68[a]

Firm	*1954*	*1955*	*1956*	*1957*	*1958*	*1959*	*1960*	*1961*	*1962*	*1963*	*1964*	*1965*	*1966*	*1967*	*1968*
AEI Semiconductors[b]	x	x	x	x	x	x	x	x	x	x	x	x	x	x	x
Ferranti	x	x	x	x	x	x	x	x	x	x	x	x	x	x	x
General Electric Company Ltd.[c]	x	x	x	x	x	x	x	x	x						
Newmarket[d]	x	x	x	x	x	x	x	x	x	x	x	x	x	x	x
Plessey[e]	x	x	x	x	x	x	x	x	x	x	x	x	x	x	x
Standard Telephones and Cables[f]	x	x	x	x	x	x	x	x	x	x	x	x	x	x	x
English Electric Valve[g]		x	x	x	x	x	x	x							
Mullard[h]		x	x	x	x	x	x	x	x	x	x	x	x	x	x
Westinghouse Brake English Electric[i]			x	x	x	x	x	x	x	x	x	x	x	x	x
Joseph Lucas[j]				x	x	x	x	x	x	x	x	x	x	x	x
Associated Transistors[k]					x	x	x	x							
Brush Clevite[l]					x	x	x	x	x	x	x	x			
MCP Electronics[m]					x	x	x	x	x	x	x				
Stone-Platt[n]					x	x	x	x							
Texas Instruments Ltd.[o]					x	x	x	x	x	x	x	x	x	x	x
International Rectifier[p]						x	x	x	x	x	x	x	x	x	x
Emihus Microcomponents[q]							x	x	x	x	x	x	x	x	x
Microwave Semiconductor Devices[r]									x						
Società Generale Semiconduttori–U.K.[s]									x	x	x	x	x	x	x
Semitron[t]										x	x	x	x	x	x
Elliott Automation Microelectronics[u]											x	x	x	x	x
Marconi Elliott Microelectronics[v]												x	x	x	x
Transitron[w]												x	x	x	x
Thorn[x]															x

Sources: Interviews with firms and other organizations with interests in the European semiconductor industry; company directories; and various articles in the trade press. Case histories for most of the British firms are found in Anthony M. Golding, "The Semiconductor Industry in Britain and the United States: A Case Study in Innovation, Growth and the Diffusion of Technology" (Ph.D. dissertation, University of Sussex, forthcoming).

a. The specific years each firm has engaged in the commercial production of semiconductors are denoted by "x." The table identifies most firms by their most recent names. Previous names and names of sister companies that also have produced semiconductors are indicated in the following notes.

b. AEI Semiconductors is part of Associated Electrical Industries (AEI), which formerly sold semiconductors through a number of other subsidiary companies (British Thomson-Houston, Siemens Edison Swan, AEI Heavy Plant Division, AEI Thorn Semiconductors) before setting up AEI Semiconductors. In 1967, General Electric Company Ltd. (see note c), which was no longer active in the semiconductor market, took over AEI. In 1968, GEC and English Electric merged; however, the English Electric subsidiary producing semiconductors, Marconi Elliott Microelectronics, and AEI Semiconductors continued to operate independently until mid-1969, and so are not consolidated in the table, which ends with 1968.

c. General Electric Company Ltd. (GEC), despite the similarity in names, is independent of the General Electric Company in the United States. In 1962, when Mullard and GEC established Associated Semiconductor Manufacturers (ASM), Mullard took over the GEC semiconductor operations. The new firm was two-thirds owned by Mullard and one-third owned by GEC until early 1969, when GEC (by then part of the GEC-English Electric combine) reduced its equity in ASM to a nominal percentage.

d. Newmarket is part of the Pye organization and in the early 1950s was called Pye Industrial Electronics. The American firm General Instrument owned 25 percent of Newmarket for a while but by 1958 the firm was wholly owned by Pye. In 1967, Philips acquired Pye, but Newmarket and Mullard have continued to operate independently.

e. Plessey began producing semiconductors (silicon alloy rectifiers) about 1954, but its first major effort was in 1957 when Semiconductors Ltd. was formed to manufacture the Philco line of transistors. This firm was owned jointly by Plessey with 51 percent of the equity and Philco (U.S.) with 49 percent until 1961, when Plessey bought out Philco.

f. A major British subsidiary of International Telephone and Telegraph (U.S.).

g. English Electric Valve stopped producing transistors in 1958, when English Electric, Ericsson Telephones, and Automatic Telephone and Electric established Associated Transistors as a joint venture (see note k), which ceased all semiconductor production in 1961.

Table 5-1, *notes (continued)*

h. The British subsidiary of Philips, the large Dutch electrical firm. See note c for information on Mullard and Associated Semiconductor Manufacturers.

i. Westinghouse Brake and Signal Company sold a 49 percent share of its semiconductor operations (excluding copper oxide and selenium rectifiers) to English Electric at the end of 1966. This joint venture is called Westinghouse Brake English Electric.

j. A large manufacturer of electrical equipment for automobiles, which produces primarily power semiconductors for its own needs.

k. Established in 1958 and owned in equal shares by English Electric, Ericsson Telephones, and Automatic Telephone and Electric, Associated Transistors produced semiconductors primarily for the in-house needs of the parent companies. This joint effort collapsed in 1961, and Associated Transistor was sold to Mullard, which in 1962 merged its semiconductor facilities with those of General Electric Company Ltd. to form Associated Semiconductor Manufacturers.

l. Brush Crystal was an affiliate of the German firm Intermetall, and imported devices from the latter. About 1958, it began small-scale semiconductor production in Britain. Its name was changed to Brush Clevite after Clevite (U.S.) acquired a majority interest in the company in 1964. In 1965, Clevite sold its semiconductor operations to ITT, and Standard Telephones and Cables (ITT's British subsidiary) absorbed Brush Clevite.

m. A subsidiary of Mining and Chemical Products Ltd., which is owned by a Swiss firm. It began producing diodes in 1958 as a joint venture with the French firm Sesco.

n. A manufacturer of marine propellers and textile machinery, which produced germanium rectifiers between 1958 and 1961.

o. A wholly owned subsidiary of Texas Instruments (U.S.).

p. Formed in 1959 by International Rectifier (U.S.) and the Lancashire Dynamo Group with each holding 50 percent of the equity. Lancashire Dynamo was later acquired by Metal Industries, which in turn was taken over by Thorn Electrical Industries in 1967.

q. Hughes (U.S.) formed Hughes International (U.K.) in 1960. In 1966, Electrical and Musical Industries (EMI) purchased a 50 percent interest and the name was changed to Emihus Microcomponents.

r. Financed by Microwave Associates (U.S.), which withdrew support in 1962 when the American electronic market was hit by a recession.

s. Società Generale Semiconduttori (SGS) is an Italian company with semiconductor operations in five European countries. Fairchild (U. S.) owned one-third interest in SGS until 1968, when it sold out to Olivetti, one of its two Italian partners. At that time, the name of the British subsidiary was changed from SGS-Fairchild.

t. Established in 1963. One of the three founders came from Plessey, another from Brush Crystal.

u. Part of Elliott-Automation. Merged with Marconi Microelectronics, forming Marconi Elliott Microelectronics in 1968, a year after the merger of its parent company with English Electric.

v. An English Electric subsidiary and now part of the General Electric Company Ltd.-English Electric combine (see note b). English Electric manufactured semiconductors in the 1950s and early 1960s through its subsidiaries English Electric Valve and Associated Transistor (see notes g and k). In the early 1960s, the production of discrete semiconductors was discontinued about the same time Marconi, another English Electric subsidiary, decided to enter the integrated circuit field. Marconi Microelectronics was established, and commercial production began about 1965. In 1968, Marconi Microelectronics and Elliott Automation Microelectronics merged, forming Marconi Elliott Microelectronics.

w. A subsidiary of Transitron (U.S.). Complete fabrication of semiconductor devices, including the diffusion of crystals, began in 1967.

x. In 1962, Associated Electrical Industries and Thorn Electrical Industries as a joint venture began to produce entertainment semiconductors for Thorn consumer equipment. When AEI and General Electric Company Ltd. merged in 1967, Thorn withdrew and began manufacturing on its own consumer semiconductors for internal use.

Société Industrielle des Liaisons Electriques (Silec), a small French manufacturer of electrical cables that began producing semiconductors in 1956 to offset stagnant conditions in the French cable market. Finally, there are the new firms like Intermetall in Germany and Semitron in Britain that were founded to engage in semiconductor manufacturing.

British, French, and German firms that engaged in the commercial production of semiconductors during the 1954–68 period and the specific years each was active in the industry are identified in Tables 5-1, 5-2, and 5-3.[4] Since the number of semiconductor firms is much smaller in

4. The tables are based primarily on information obtained in interviews with managers of semiconductor firms, trade association personnel, government officials,

Table 5-2. Semiconductor Firms in France and Years They Were Active in the Industry, 1954–68[a]

Firm	1954	1955	1956	1957	1958	1959	1960	1961	1962	1963	1964	1965	1966	1967	1968
Compagnie des Dispositifs Semi-conducteurs Westinghouse (CDSW)[b]	x	x	x	x	x	x	x	x	x	x	x	x	x	x	x
Compagnie Générale d'Electricité (CGE)[c]	x	x	x	x	x	x	x	x	x	x	x	x	x	x	x
Compagnie Générale des Semi-conducteurs (Cosem)[d]	x	x	x	x	x	x	x	x	x	x	x	x	x	x	x
Le Matériel Téléphonique (LMT)[e]	x	x	x	x	x	x	x	x	x	x	x	x	x	x	x
Lignes Télégraphiques et Téléphoniques (LTT)[f]	x	x	x	x	x	x	x	x	x	x	x	x	x	x	x
Radiotechnique Compelec (RTC)[g]	x	x	x	x	x	x	x	x	x	x	x	x	x	x	x
Sescosem[h]	x	x	x	x	x	x	x	x	x	x	x	x	x	x	x
Soral[i]	x	x	x	x	x	x	x	x	x	x	x	x	x	x	x
Société Industrielle de Liaisons Electriques (Silec)[j]			x	x	x	x	x	x	x	x	x	x	x	x	x
Compagnie Industrielle pour la Transformation de l'Energie (Cogie)[k]							x	x	x	x	x	x	x	x	x
Texas Instruments–France[l]							x	x	x	x	x	x	x	x	x
International Business Machines–France[m]										x	x	x	x	x	x
Società Generale Semiconduttori–France[n]												x	x	x	x
Motorola[o]															x
Transitron[p]															x
Sprague-France[q]															x

Sources: Interviews with firms and other organizations with interests in the European semiconductor industry; company directories; and various articles in the trade press.

a. The specific years each firm has engaged in the commercial production of semiconductors are denoted by "x." The table identifies most firms by their most recent name. Names of predecessor companies, former names under which semiconductors were sold, or names of sister companies which have also produced semiconductors are indicated in the following notes.

b. Westinghouse Electric (U.S.) bought the rectifier division of Compagnie des Freins et Signaux Westinghouse and joined with Société Jeumont-Schneider in 1967 to form CDSW, of which Westinghouse owns 82 percent. From the early 1950s until Westinghouse Electric bought its rectifier division, the Compagnie des Freins et Signaux Westinghouse produced semiconductors for the French market.

c. CGE has concentrated its semiconductor activities, which are limited and confined primarily to power devices, at its Marcoussis research laboratory. In the early 1960s it did establish Compelec (Compagnie Générale des Composants Electroniques) to produce semiconductors. In 1967, as part of a larger agreement with Philips (Netherlands), Compelec was severed from CGE and absorbed by Radiotechnique. CGE has since maintained a small effort that includes some production at Marcoussis.

d. Compagnie Générale de Télégraphie Sans Fil (CSF) established Cosem in the early 1960s to assume its semiconductor activities. Before that time, semiconductors were sold under the CSF name. For a period of about two years ending in 1967, Cosem and Silec (see note j) merged their marketing activities and both sold devices under the name of Cosil. In 1968, Cosem merged with Sesco, a subsidiary of Compagnie Française Thomson-Houston (CFTH), and formed Sescosem. The two parent companies, CSF and CFTH had merged a year earlier.

e. A subsidiary of International Telephone and Telegraph (U.S.). In recent years, some of the semiconductor activities previously performed by LMT have been shifted to Intermetall, a German subsidiary of ITT that has a plant at Colmar, France. Laboratorie Central de Télécommunications (LCT), a sister company, is primarily engaged in research activities. It is one of the major research facilities serving the whole ITT system.

f. LTT is owned in roughly equal shares by ITT (U.S.) and two French firms. Its production of semiconductors is very small.

g. Radiotechnique Compelec is a subsidiary of the Dutch firm Philips. Before absorbing Compelec in 1967 (see note c), it was called Radiotechnique.

h. Compagnie Française Thomson-Houston (CFTH) created the Société Européenne des Semiconducteurs (Sesco) to assume its semiconductor operations in the early 1960s. Previously, CFTH had marketed semiconductors under its own name. General Electric (U.S.) participated in the formation of Sesco and acquired

Table 5-2, *notes (continued)*

a 49 percent interest in the firm, but was forced to relinquish this equity when Sesco and Cosem merged in 1968, forming Sescosem, an all-French semiconductor firm.

i. A small firm specializing in low-power devices.

j. Silec began semiconductor production in 1956 and has concentrated on diodes and rectifiers.

k. Cogie is a very small semiconductor producer specializing in power devices. It is a subsidiary of Ferrodo, which manufactures automobile parts. The precise date it began semiconductor production is unknown.

l. A wholly owned subsidiary of Texas Instruments (U.S.).

m. IBM-France manufactures semiconductors for IBM computer facilities throughout Europe. Semiconductor production is solely for in-house use.

n. A subsidiary of Società Generale Semiconduttori (Italy). Fairchild (U.S.) held a one-third interest in the parent company until 1968, at which time the name of the firm was changed from SGS-Fairchild.

o. A subsidiary of Motorola (U.S.).

p. A subsidiary of Transitron (U.S.).

q. A subsidiary of Sprague (U.S.).

Europe than in America, the tables include all semiconductor firms rather than just transistor producers. If just the latter were considered, a good number of the new firms, which specialize in diodes and rectifiers, would be excluded.

The tables illustrate a number of interesting features in the evolution of the European semiconductor industry. First, the receiving tube producers in Britain, France, and Germany, as in America, entered the semiconductor industry early. Nearly all were engaged in commercial production by 1954.

Second, the number of new European firms entering the industry has been small, particularly in comparison with the multitude of new American firms. Within Europe, England has the greatest number of new firms and Germany the fewest. Variation in market size and the possibility that the tables have omitted a few new European firms may account for the intra-European differences, but it is unlikely that these factors fully explain the much larger difference between the European and American industries.

Third, the barriers inhibiting the entry of new European firms apparently are increasing. Many were already producing semiconductors in 1954, but very few entered the semiconductor industry in the 1960s.

Fourth, entry barriers for foreign subsidiaries appear to be moving in the opposite direction. Few American firms established semiconductor facilities in Europe during the 1950s. The first was Texas Instruments, which set up its British subsidiary in 1957 aand began production the following year. Throughout the 1960s, however, numerous American

---

and members of universities and research organizations with interests in the semiconductor industry. Trade directories and articles from the trade press were used to supplement this information. However, some of the evidence was conflicting, and a few firms may have been overlooked, particularly small firms that produced semiconductors for only a short time.

Table 5-3. Semiconductor Firms in Germany and Years They Were Active in the Industry, 1954–68[a]

Firm	*1954*	*1955*	*1956*	*1957*	*1958*	*1959*	*1960*	*1961*	*1962*	*1963*	*1964*	*1965*	*1966*	*1967*	*1968*
Allgemeine Elektrizitäts Gesellschaft–Telefunken[b]	x	x	x	x	x	x	x	x	x	x	x	x	x	x	x
Intermetall[c]	x	x	x	x	x	x	x	x	x	x	x	x	x	x	x
Siemens	x	x	x	x	x	x	x	x	x	x	x	x	x	x	x
Standard Elektrik Lorenz[d]	x	x	x	x	x	x	x	x	x	x	x	x	x		
Valvo[e]			x	x	x	x	x	x	x	x	x	x	x	x	x
Eberle						x	x	x	x	x	x	x	x	x	x
Semikron							x	x	x	x	x	x	x	x	x
Società Generale Semiconduttori–Deutschland[f]													x	x	x
Texas Instruments–Deutschland[g]													x	x	x

Sources: Interviews with firms and other organizations with interests in the European semiconductor industry; firm directories; and various articles in the trade press.

a. The specific years each firm has engaged in the commercial production of semiconductors are denoted by "x." The table identifies firms by their most recent names. Previous names are indicated in the following notes.

b. A wholly owned subsidiary of Allgemeine Elektrizitäts Gesellschaft (AEG).

c. Founded in 1952, acquired by Clevite (U.S.) in 1955, and sold with the rest of Clevite's semiconductor operations to International Telephone and Telegraph (U.S.) in 1965.

d. A major German subsidiary of ITT (U.S.). In the early 1950s, it sold semiconductors under the plant name Sueddeutsche Apparate Fabrik. Standard Elektrik Lorenz is not listed as a semiconductor producer after 1966 because ITT began selling its German semiconductors under the Intermetall trademark shortly after acquiring the firm.

e. The German subsidiary of the Dutch firm Philips. The date Valvo began producing semiconductors in Germany is difficult to determine because the firm may have relied for a time on imports from Radiotechnique, Mullard, or other Philips subsidiaries.

f. A subsidiary of Società Generale Semiconduttori (Italy). Fairchild owned one-third interest in the parent company until 1968, at which time the name of the firm was changed from SGS-Fairchild.

g. A wholly owned subsidiary of Texas Instruments (U.S.).

firms followed Texas Instruments to Europe, and there is little indication that this movement is about to subside.[5] Motorola, General Instrument, and Union Carbide are building new plants in Britain. Fairchild, having severed its ties with SGS, is constructing a wholly owned production facility in Germany. Both Signetics and National Semiconductor are setting up new facilities in Britain as well as Germany.

Fifth, very few firms in Europe have abandoned semiconductor production, which is in marked contrast to the American industry. In France and Germany, the mortality rate is almost zero; Cosem and Standard Elektrik Lorenz are no longer independent semiconductor producers, but both firms were merely consolidated with other semiconductor firms.

5. The increased migration of American firms to Europe in the sixties, described here for the semiconductor industry, occurred in other industries as well. This general movement was stimulated by the creation of the European Economic Community and the relaxation of currency restrictions in Europe.

Contributions by Firms

In Europe, as in the United States, semiconductor firms have varied greatly in their R&D, production, and marketing performance. Some, despite substantial research efforts, have not been able to increase or even maintain their share of the semiconductor market, while others, with little research, have, primarily by swiftly introducing into their production lines the new technology developed in their own and other laboratories. This section examines the contributions to the innovative and diffusion processes of firms producing semiconductors in Britain, France, and Germany, and assesses the relative strengths of the receiving tube producers, the new European firms, and the new foreign subsidiaries in each of these activities.

Innovative Process

Since information on semiconductor R&D expenditures by European firms is sparse,[6] this section relies on semiconductor patents and major innovations to assess firms' contributions to the innovative process.

SEMICONDUCTOR PATENTS. Tables 5-4, 5-5, and 5-6 indicate the number of French semiconductor patents acquired over the 1954–68 period by the semiconductor firms of Britain, France, and Germany. Two characteristics of these data should be noted before their implications for the innovative process are assessed. First, the interval between application and award generally is much shorter than for the American patents considered in the previous chapter. This lag averages about a year, and seldom exceeds two years. French patents, therefore, reflect company

6. In England, it is known that Mullard and the GEC–English Electric combine (which includes AEI, Marconi Elliott, and GEC) are spending the most on semiconductor R&D, followed by Plessey, STC (including Standard Telecommunication Laboratories), and Ferranti. See Anthony M. Golding, "The Semiconductor Industry in Britain and the United States: A Case Study in Innovation, Growth and the Diffusion of Technology" (Ph.D. dissertation, University of Sussex, forthcoming).

The receiving tube firms were also among the first firms in Europe to undertake semiconductor R&D. By the early 1950s, Philips, ITT, AEI, CSF, CFTH, and Siemens were actively engaged in semiconductor research. See C. C. Gee, "World Trends in Semiconductor Development and Production," *British Communications and Electronics* (June 1959).

R&D activity a year or two earlier, rather than three or four years earlier as do American patents.[7]

Second, the patents recorded for Philips, Texas Instruments, and other international firms come from all of their operations and consequently reflect the contribution to the innovative process of the entire firm rather than just the subsidiary operating in England, France, or Germany. This procedure was followed because the subsidiaries of these companies draw on the R&D activities of the entire firm in competing for the semiconductor business of European countries.

Among the various types of companies producing semiconductors in Europe, the table shows that the receiving tube firms have acquired by far the greatest number of patents. These firms particularly dominate in Germany, accounting for 88 percent of the semiconductor patents awarded during 1954–68. In Britain, where new firms and foreign subsidiaries are most numerous, the receiving tube firms have acquired twice as many patents as the other two groups together. Even in France, where their relative performance is poorest, they hold more than half the total patent awards.

7. Several other differences between French and American patents should also be noted. First, because firms have a greater propensity to patent in their own countries, French patents tend to overstate the contributions to the innovative process of the receiving tube producers and new firms in France, though the bias against foreign subsidiaries should not be great, since the latter have French manufacturing facilities and hence strong incentives to patent there too. Conversely, American patents would favor American subsidiaries. When French patents are used to evaluate the contributions of the semiconductor firms of Britain or Germany, the biases are probably less. Yet some remain, for a few firms like Philips, ITT, and Texas Instruments operate in France as well as Britain and Germany and so have a greater propensity to acquire French patents than their competitors in the other two countries.

Second, the requirements, such as demonstrating originality, are less stringent for a French patent than for an American patent. Typically, a questionable patent is challenged in court after it is awarded, which explains the much shorter patent pendency period in France. Finally, the patent classification system varies between the two countries and, in turn, in the coverage of semiconductor patents. The French patents, for example, do not include new semiconductor applications in final electronic products, while the American patents do.

To evaluate the importance of the special characteristics and biases associated with French patents, semiconductor patents issued by the United States and (for the semiconductor firms of England) by Britain were also examined. (The British patents were collected by Anthony M. Golding from *Patents for Inventions: Abridgements of Specifications* [London: Her Majesty's Stationery Office, relevant years]). These patents support the conclusions reached with French patents.

Table 5-4. French Semiconductor Patents Awarded to Firms in Great Britain, 1954–68

Type and name of firm[a]	*1954*	*1955*	*1956*	*1957*	*1958*	*1959*	*1960*	*1961*	*1962*	*1963*	*1964*	*1965*	*1966*	*1967*	*1968*	*Total 1954 –68*
Receiving tube firms																
Associated Electrical Industries	0	1	0	1	6	17	17	10	12	17	8	7	6	3	9	114
Standard Telephones and Cables	7	12	4	8	1	15	2	5	7	9	11	6	13	13	24	137
Mullard	7	10	15	8	22	38	41	41	40	43	26	34	42	23	68	458
Subtotal	**14**	**23**	**19**	**17**	**29**	**70**	**60**	**56**	**59**	**69**	**45**	**47**	**61**	**39**	**101**	**709**
New firms																
General Electric Company Ltd.	2	0	6	3	9	7	3	2	7	4	1	0	0	0	0	44
Joseph Lucas	0	0	0	0	0	0	0	0	1	1	0	2	3	3	1	11
Marconi Elliott Micro-electronics	0	1	0	1	1	2	2	0	1	0	0	1	0	0	0	9
Newmarket	0	0	1	1	0	2	0	0	1	0	0	0	0	0	0	5
Plessey	0	0	0	0	0	6	0	0	0	0	1	3	0	0	1	11
Westinghouse Brake and Signal	0	0	3	11	5	3	0	4	10	7	5	1	7	2	2	60
Associated Transistors	0	0	0	0	0	0	0	0	2	0	0	0	0	0	0	2
Subtotal	**2**	**1**	**10**	**16**	**15**	**20**	**5**	**6**	**22**	**12**	**7**	**7**	**10**	**5**	**4**	**142**
New foreign subsidiaries																
Texas Instruments Ltd.	0	0	0	0	7	0	8	7	15	11	3	10	22	18	21	122
Brush Clevite	0	0	0	0	0	0	4	2	4	5	3	1	0	0	0	19
International Rectifier	0	0	0	0	0	0	1	0	0	1	2	2	1	2	9	18
Emihus Microcomponents	0	0	0	0	2	7	1	2	4	3	0	3	4	1	1	28
Società Generale Semiconduttori-U.K.	0	0	0	0	0	0	0	6	2	0	4	1	2	1	2	18
Subtotal	**0**	**0**	**0**	**0**	**9**	**7**	**14**	**17**	**25**	**20**	**12**	**17**	**29**	**22**	**33**	**205**
Total	**16**	**24**	**29**	**33**	**53**	**97**	**79**	**79**	**106**	**101**	**64**	**71**	**100**	**66**	**138**	**1,056**
Percentage of total																
Receiving tube firms	*87*	*96*	*66*	*52*	*55*	*72*	*76*	*71*	*56*	*68*	*70*	*66*	*61*	*59*	*73*	*67*
New firms	*13*	*4*	*34*	*48*	*28*	*21*	*6*	*8*	*21*	*12*	*11*	*10*	*10*	*8*	*3*	*14*
New subsidiaries	*0*	*0*	*0*	*0*	*17*	*7*	*18*	*21*	*23*	*20*	*19*	*24*	*29*	*33*	*24*	*19*

Sources: France, Institut National de la Propriété Industrielle, *Brevets d'Invention: Table des Brevets et Certificats d'Addition Imprimés*, Vol. 2: *Table des Brevets par Ordre des Matières* (Paris: Imprimerie Nationale, annual). Semiconductor patents awarded before 1959 were filed under the French patent category, "Transport et Mesure d'Electricité, Appareils Divers," and were identified by their title. Since 1959, semiconductor patents have had their own category: "Dispositifs à Semi-Conducteurs."

a. Firms identified in Table 5-1 that are not listed in this table received no French semiconductor patents during 1954–68. Patents attributed to individual firms include those awarded to parent firms and sister divisions as well as those granted under earlier firm names.

Among the receiving tube firms, the Philips subsidiaries—Mullard and RTC—have the largest patent holdings in England and France. In Germany, Siemens is the leader. Except for Telefunken in Germany and CGE in France, all the receiving tube firms have been accumulating semiconductor patents since the early or middle fifties.

The new foreign subsidiaries have acquired only a minor share of semiconductor patents—19 percent of the British total, 32 percent of the

Table 5-5. French Semiconductor Patents Awarded to Firms in France, 1954–68

Type and name of firm[a]	*1954*	*1955*	*1956*	*1957*	*1958*	*1959*	*1960*	*1961*	*1962*	*1963*	*1964*	*1965*	*1966*	*1967*	*1968*	*Total 1954–68*
Receiving tube firms																
Compagnie Française Thomson-Houston	24	20	14	15	18	19	12	14	21	10	7	5	9	2	16	206
Compagnie Générale d'Electricité	0	0	0	0	0	0	0	1	2	1	4	3	4	3	7	25
Compagnie Générale de Télégraphie Sans Fil	2	3	1	2	6	4	1	2	5	9	3	3	2	5	0	48
Radiotechnique Compelec	7	10	15	8	22	38	41	41	40	43	26	34	42	23	68	458
Subtotal	**33**	**33**	**30**	**25**	**46**	**61**	**54**	**58**	**68**	**63**	**40**	**45**	**57**	**33**	**91**	**737**
New firms																
Compagnie des Dispostifs Semiconducteurs Westinghouse	1	0	2	1	1	1	1	1	0	2	0	1	0	0	0	11
Le Matériel Téléphonique[b]	7	12	4	8	1	15	2	5	7	9	11	6	13	13	24	137
Société Industrielle de Liaisons Electriques	0	0	0	0	0	3	0	1	0	1	0	1	1	0	3	10
Soral	0	0	0	0	0	0	1	0	1	0	0	0	0	0	0	2
Subtotal	**8**	**12**	**6**	**9**	**2**	**19**	**4**	**7**	**8**	**12**	**11**	**8**	**14**	**13**	**27**	**160**
New foreign subsidiaries																
International Business Machines–France	0	1	4	2	18	4	5	9	12	13	27	29	38	27	37	226
Motorola	0	0	0	0	0	0	0	0	1	5	8	4	13	7	13	51
Società Generale Semiconduttori–France	0	0	0	0	0	0	0	6	2	0	4	1	2	1	2	18
Sprague-France	0	0	0	0	0	0	0	0	0	0	0	0	3	0	3	6
Texas Instruments–France	0	0	0	0	7	0	8	7	15	11	3	10	22	18	21	122
Subtotal	**0**	**1**	**4**	**2**	**25**	**4**	**13**	**22**	**30**	**29**	**42**	**44**	**78**	**53**	**76**	**423**
Total	**41**	**46**	**40**	**36**	**73**	**84**	**71**	**87**	**106**	**104**	**93**	**97**	**149**	**99**	**194**	**1,320**
Percentage of total																
Receiving tube firms	*80*	*72*	*75*	*69*	*63*	*72*	*76*	*67*	*64*	*61*	*43*	*47*	*38*	*33*	*47*	*56*
New firms	*20*	*26*	*15*	*25*	*3*	*23*	*6*	*8*	*8*	*11*	*12*	*8*	*9*	*13*	*14*	*12*
New foreign subsidiaries	*0*	*2*	*10*	*6*	*34*	*5*	*18*	*25*	*28*	*28*	*45*	*45*	*53*	*54*	*39*	*32*

Sources: France, Institut National de la Propriété Industrielle, *Brevets d'Invention: Table des Brevets et Certificats d'Addition Imprimés*, Vol. 2: *Table des Brevets par Ordre des Matières* (Paris: Imprimerie Nationale, annual). Semiconductor patents awarded before 1959 were filed under the French patent category, "Transport et Mesure d'Electricité, Appareils Divers," and were identified by their title. Since 1959, semiconductor patents have had their own category: "Dispositifs à Semi-Conducteurs."

a. Firms identified in Table 5-2 that are not listed in this table received no French semiconductor patents during 1954–68. Patents attributed to individual firms include those awarded to parent firms and sister divisions as well as those granted under earlier company names.

b. Produces industrial tubes but not receiving tubes; has been an International Telephone and Telegraph subsidiary since World War II. For these reasons, the firm is not included with the receiving tube producers or the new foreign subsidiaries. As an American subsidiary, however, it is not typical of the new French firms either. If excluded from this group, the latter's share of patents would be even less significant.

French total, and 10 percent of the German total. The larger share attributed to this group in France is due primarily to IBM-France, which is not a typical new foreign subsidiary.

New European firms have generated even fewer semiconductor pat-

Table 5-6. French Semiconductor Patents Awarded to Firms in Germany, 1954–68

Type and name of firm[a]	*1954*	*1955*	*1956*	*1957*	*1958*	*1959*	*1960*	*1961*	*1962*	*1963*	*1964*	*1965*	*1966*	*1967*	*1968*	*Total 1954 -68*
Receiving tube firms																
Siemens	16	9	10	4	17	36	53	41	45	52	58	69	103	41	47	601
Standard Elektrik Lorenz	7	12	4	8	1	15	2	5	7	9	11	6	13	13	24	137
Allgemeine Elektrizitäts Gesellschaft–Telefunken	0	0	0	0	0	0	2	6	7	2	2	8	17	10	10	64
Valvo	7	10	15	8	22	38	41	41	40	43	26	34	42	23	68	458
Subtotal	**30**	**31**	**29**	**20**	**40**	**89**	**98**	**93**	**99**	**106**	**97**	**117**	**175**	**87**	**149**	**1,260**
New firms																
Intermetall	0	0	0	0	0	0	3	5	3	4	3	4	0	0	0	22
Semikron	0	0	0	0	0	0	0	0	1	0	0	0	1	0	0	2
Subtotal	**0**	**0**	**0**	**0**	**0**	**0**	**3**	**5**	**4**	**4**	**3**	**4**	**1**	**0**	**0**	**24**
New foreign subsidiaries																
Società Generale Semiconduttori–Deutschland	0	0	0	0	0	0	0	6	2	0	4	1	2	1	2	18
Texas Instruments–Deutschland	0	0	0	0	7	0	8	7	15	11	3	10	22	18	21	122
Subtotal	**0**	**0**	**0**	**0**	**7**	**0**	**8**	**13**	**17**	**11**	**7**	**11**	**24**	**19**	**23**	**140**
Total	**30**	**31**	**29**	**20**	**47**	**89**	**109**	**111**	**120**	**121**	**107**	**132**	**200**	**106**	**172**	**1,424**
Percentage of total																
Receiving tube firms	*100*	*100*	*100*	*100*	*85*	*100*	*90*	*84*	*83*	*88*	*91*	*89*	*87*	*82*	*87*	*88*
New firms	*0*	*0*	*0*	*0*	*0*	*0*	*3*	*4*	*3*	*3*	*3*	*3*	*1*	*0*	*0*	*2*
New foreign subsidiaries	*0*	*0*	*0*	*0*	*15*	*0*	*7*	*12*	*14*	*9*	*6*	*8*	*12*	*18*	*13*	*10*

Sources: France, Institut National de la Propriété Industrielle, *Brevets d'Invention: Table des Brevets et Certificats d'Addition Imprimés*, Vol. 2: *Table des Brevets par Ordre des Matières* (Paris: Imprimerie Nationale, annual). Semiconductor patents awarded before 1959 were filed under the French patent category, "Transport et Mesure d'Electricité, Appareils Divers," and were identified by their title. Since 1959, semiconductor patents have had their own category: "Dispositifs à Semi-Conducteurs."

a. Firms identified in Table 5-3 that are not listed in this table received no French semiconductor patents during 1954–68. Patents attributed to individual firms include those awarded to parent firms and sister divisions as well as those granted under earlier firm names.

ents than new subsidiaries. The number obtained by the new German firms is negligible. This is also true for the new French firms if LMT is excluded from this group because of its ties with ITT and the latter's receiving tube affiliates in England and Germany.

The evolution of patent shares over the 1954–68 period is also shown in Tables 5-4 to 5-6. While the proportion of patents acquired by the new European firms has remained stable or even declined slightly over time, the share of the new foreign subsidiaries has risen. Receiving tube firms no longer account for the very high percentage of semiconductor patents as in the 1950s, though they still generate more new patents than either of the other groups.

These findings suggest that among the European semiconductor firms the receiving tube firms have made the greatest contribution to the innovative process. Consequently, in competing for the European market, they should not be at a disadvantage because of an inadequate level of innovative activity.[8]

MAJOR INNOVATIONS. Another measure of company contributions to the innovative process is the number of major innovations introduced. According to Tables 2-1 and 2-2, American firms with new subsidiaries in Europe account for five and a half of the major product innovations and one of the major process innovations. The European receiving tube firms are credited with only one major innovation—3-5 compounds introduced by Siemens—and the new European firms with none.

To some extent, Tables 2-1 and 2-2 understate the contribution of the European receiving tube firms by attributing the Gunn diode to IBM. While IBM made the first laboratory model, Associated Semiconductor Manufacturers, controlled by Mullard, first made Gunn diodes commercially. In addition, as the earlier discussion of the tables noted, a few innovations that some consider major were not included. Two of these, the germanium power rectifier and the alloy diffused transistor, were introduced by European receiving tube firms.

Despite these qualifications, the receiving tube firms in Europe have not produced as many major semiconductor innovations as the large number of patents they have acquired implies. This is consistent with the common assertion that good research in Europe often is not fully exploited because development proceeds too slowly or is otherwise inadequate.[9]

8. These results do not necessarily imply that the innovative performance of Britain, France, and Germany would improve if there were more or bigger receiving tube firms in the European semiconductor industry and fewer firms of other types. For one thing, patent shares have not been adjusted for differences in market shares. (The next section, however, indicates that in the 1960s the new foreign subsidiaries captured a share of the European semiconductor market that exceeded their share of patents. If worldwide semiconductor sales are considered, as total patents are, the disparities in the patent and market shares obtained by the two types of firms are even greater.) For another, several important receiving tube firms carry on research and related activities in numerous countries, and many of the advances they develop do not occur in Britain, France, or Germany. (On the other hand, the same is also true for the new foreign subsidiaries, which with few exceptions depend almost entirely on the research of parent firms in the United States.) Finally, other factors such as the fact that new firms and foreign subsidiaries may stimulate the innovative activity of the receiving tube producers should be taken into account.

9. See, for example, Christopher Freeman, "Research and Development in Electronic Capital Goods," *National Institute Economic Review,* No. 34 (November 1965), pp. 67–70.

Before turning to company performance in the diffusion process, it should be noted that many semiconductor patents and most of the major innovations were produced by American firms, in particular Bell Laboratories, that have no manufacturing facilities in Europe. As a result, the competitive positions of the various types of firms operating in Europe have largely been determined by their ability to adopt the advances in technology developed by other firms.

Diffusion Process

The diffusion of new semiconductor technology in Europe can be separated into two steps or stages. The first, necessary because so many new developments originated in the United States, involves the intercountry transfer of innovations. The second covers the intracountry diffusion that follows once the technology is successfully implanted in the country.

Table 5-7 provides some insights into company contributions to the intercountry transfer of innovations. It shows that the receiving tube firms have initiated the commercial production of more major semiconductor devices in Europe than either the new firms or the new foreign subsidiaries. They have completely dominated in France, though in England and Germany new firms have also contributed appreciably to this stage of diffusion.

Only in England have new foreign subsidiaries been the first to produce any significant number of the important semiconductor devices. The small size of their contribution is due in part to their late arrival on the European scene. No new foreign subsidiaries existed in Britain until 1957, in France until 1960, and in Germany until 1966.

Table 5-7. Percentage of Major Semiconductor Devices First Produced in Great Britain, France, and Germany, by Type of Firm, and Average Imitation Lags, 1954–68

	Percentage of innovations initiated[a]			*Average imitation lag (years)*		
Type of firm	*Britain*	*France*	*Germany*	*Britain*	*France*	*Germany*
Receiving tube firms	*47*	*96*	*76*	1.9	2.7	2.6
New firms	*32*	*0*	*24*	2.1	...	2.0
New foreign subsidiaries	*21*	*4*	*0*	2.4	3.0	...

Source: Table 3-1.

a. When two or three firms are responsible for the first production of a major semiconductor device in a country, each is given one-half or one-third the credit.

The average imitation lags associated with the three types of companies, also shown in Table 5-7, vary from country to country. Receiving tube firms have a shorter lag than new domestic firms in England, but not in Germany. The lag for the new foreign subsidiaries in England is longer than the lags for the English receiving tube and new firms, but it would be shorter than both if the silicon transistor were omitted (see Table 3-1). Texas Instruments initiated the production of this device in England in 1958, the first year the company manufactured in England but four years after silicon transistors were first introduced in the United States. Although the data are far too sparse to draw any solid conclusions, they are consistent with the hypothesis that foreign subsidiaries, once established, provide a relatively swift channel for the intercountry diffusion of new innovations, but that firms without foreign subsidiaries may delay diffusion if they try to capitalize abroad on innovations by establishing subsidiaries rather than licensing existing foreign firms.

To assess company contributions toward intracountry diffusion, we rely as in the previous chapter on market-share data. Again, firms that capture a share of the market exceeding their contribution to the innovative process are assumed to do so by using the new techniques generated in their own and other research facilities to produce better and cheaper semiconductors.

Precise market-share data are not available for earlier years, although it is known that receiving tube firms dominated the British, French, and German semiconductor markets in the 1950s. New firms produced mainly special devices in limited quantities. New foreign subsidiaries, with an exception or two toward the end of the decade, were not yet active in Europe, and only a small part of the demand was filled by imports from the United States.

Among the receiving tube firms, Philips and its subsidiaries—Mullard, Radiotechnique, and Valvo—were especially strong. Philips developed during the 1950s a good line of devices, which included the germanium alloy junction transistor, and by 1960 probably accounted for over half of all semiconductor sales in England, France, and Germany, particularly dominating the consumer market.

During the 1960s, the receiving tube firms lost a sizable segment of the market to new foreign subsidiaries. By 1968, as Table 5-8 indicates, new subsidiaries supplied nearly a quarter of the German market, a third of the French market, and about half of the English market. Texas Instruments, SGS with Fairchild assistance, and Motorola have made the greatest inroads into the European markets.

Table 5-8. Semiconductor Market Shares in Great Britain, France, and Germany, by Firm, 1968

Great Britain		*France*		*Germany*	
Firm	*Share*	*Firm*	*Share*	*Firm*	*Share*
Receiving tube firms		*Receiving tube firms*[a]		*Receiving tube firms*	
Mullard	22	Radiotechnique Compelec	22	Valvo	25
Standard Telephones & Cables	6	Sescosem	20	Siemens	22
Associated Electrical Industries	4			Intermetall[b]	10
				Allgemeine Elektrizitäts Gesellschaft–Telefunken	9
Subtotal	**32**	**Subtotal**	**42**	**Subtotal**	**66**
New firms		*New firms*		*New firms*	
Ferranti	5	Silec	7	Semikron	1
Westinghouse Brake English Electric	5	Others	3	Eberle	1
Marconi Elliott Microelectronics	3				
Others	3				
Subtotal	**16**	**Subtotal**	**10**	**Subtotal**	**2**
New foreign subsidiaries		*New foreign subsidiaries*[c]		*New foreign subsidiaries*	
Texas Instruments Ltd.	23	Texas Instruments–France	20	Texas Instruments–Deutschland	16
Società Generale Semiconduttori–U.K.	14	Società Generale Semiconduttori–France	7	Società Generale Semiconduttori–Deutschland	6
International Rectifier	3	Motorola	5		
Emihus Microcomponents	2	Transitron	1		
Transitron	2				
Subtotal	**44**	**Subtotal**	**33**	**Subtotal**	**22**
Importers[d]		*Importers*[d]		*Importers*[d]	
Motorola	5	Intermetall	3	Motorola	4
Others	3	Others	12	Others	6
Subtotal	**8**	**Subtotal**	**15**	**Subtotal**	**10**
Total	**100**	**Total**	**100**	**Total**	**100**

Sources: Interviews with managers of semiconductor firms, trade association personnel, and government officials. The British figures are based primarily on data found in Golding, "The Semiconductor Industry in Britain and the United States."

a. Compagnie Générale d'Electricité is not included among the receiving tube firms because its limited semiconductor production at its Marcoussis laboratory is a negligible part of the overall French market.

b. Since 1965, an International Telephone and Telegraph subsidiary and an affiliate of Standard Elektrik Lorenz, a German receiving tube producer. Consequently, it is included with receiving tube firms.

c. International Business Machines is not included. Its semiconductor production is all for its own use; since this is substantial, the share of the French semiconductor market held by the new foreign subsidiaries would be appreciably larger if IBM were included.

d. Only firms without domestic manufacturing facilities are considered importers. Such firms account for only part of total imports since new foreign subsidiaries and other semiconductor manufacturers import devices to supplement domestic production.

New firms, though in the market long before the new foreign subsidiaries, have not increased their market share during the 1960s. Except for Ferranti and more recently Marconi Elliott Microelectronics, these firms have specialized and not ventured into the general high-volume semiconductor markets. Silec, for example, has concentrated on diodes and rectifiers since entering the industry in 1956. It produces only a few special types of transistors even though it was one of the first French firms to use silicon, the material now used with the planar process to batch-fabricate transistors and integrated circuits. Similarly, Intermetall, at least until acquired by ITT, hesitated to enter the mass markets and directly challenge Siemens and Valvo despite an early lead over these giants in the planar techniques. Other new European firms confining their semiconductor operations to limited and specialized markets include WBEE, Joseph Lucas, Semitron, Cogie, Soral, Eberle, and Semikron.

One searches in vain for a new European firm which, after introducing a successful device or two, expands its line to encompass a wide range of high-volume devices and thereby eventually captures a sizable share of the market. There are no firms in this group comparable to Texas Instruments, Transitron, Fairchild, and Motorola in the American market. The closest counterparts to the latter in Europe are the new foreign firms—subsidiaries of the successful new firms in the United States.

There are, to be sure, differences between the behavior and performance of the new subsidiaries in Europe and the new firms in the United States. First, the former captured a large share of the European market only in the 1960s, while new firms have provided the sales leaders in the United States since the mid-1950s. Second, new American firms have successfully competed in almost all types of semiconductors except power devices. In Europe, receiving tube firms dominate the production of consumer devices, a large part of the European market, as well as of power devices.

Despite the differences, the new subsidiaries in Europe and the new companies in America are similar in many respects. Compared with their competition, both are particularly strong in production and marketing activities. Both are quick to introduce new developments into their production lines. Both are adept at volume production that promotes low prices and mass markets.

The new foreign subsidiaries like the new firms in America have quickly exploited market opportunities that arise when established companies lag in introducing new techniques and devices. They were able to penetrate the European markets in the sixties primarily because the receiving tube firms were slow in adopting silicon and the related technology essential for the batch-fabrication of discrete semiconductors as well as integrated circuits.

Like Philco in the United States, some European receiving tube firms were victims of their own success. Philips, in particular, did very well with its germanium alloy junction and post alloy diffused transistors in the fifties and early sixties, and was naturally reluctant to switch to a new and radically different technology. Consequently, the firm was late in using silicon planar techniques: volume production of planar transistors and integrated circuits began about 1966 and 1967, respectively, several years behind Texas Instruments and SGS-Fairchild.

Since new foreign subsidiaries have concentrated in those areas, usually the most sophisticated technologically, that the receiving tube firms have been slow to enter, their market share for many of the latest devices, such as integrated circuits, far exceeds their share of the overall market. In winning these advanced markets, the new foreign subsidiaries have assured European equipment firms access to new devices that the established semiconductor firms were slow in providing. Moreover, by challenging the receiving tube firms in their home market, they have stimulated the latter to respond more quickly to new technology.

Despite the economic benefits these firms have brought, there is much concern in Europe that a large part of the semiconductor market and particularly the more sophisticated sectors are dominated by companies controlled by foreign interests. This raises the question, Why have the new European firms not behaved or performed as have the new firms in the United States? If they had, there would have been less need and opportunity for foreign firms in Europe. The rest of this chapter seeks the reasons why the barriers to entry into the large-volume semiconductor markets are high for new European firms but low for new foreign subsidiaries.

Availability of Technology

In the United States, new firms can acquire state-of-the-art technology in the semiconductor field because innovating firms follow liberal licensing policies and because scientists and engineers frequently move from one company to another. This section examines the licensing policies and the mobility of scientists and engineers in the European semiconductor industry to determine how they affect the entry barriers for new firms and foreign subsidiaries.

Licenses and Patents

A few American firms have produced most of the major semiconductor innovations and hold a substantial share of the important patents. It is primarily to these companies that the European firms, like other American firms, have turned for licenses. During the 1950s, a Western Electric license covering the strategic Bell patents sufficed for most European firms, though a few also sought agreements with RCA, General Electric, Westinghouse, and Philco. With the rise of planar technology and integrated circuits in the 1960s, European firms found a Fairchild or Texas Instruments license desirable as well.

In assessing the effect of licensing on entry conditions, two functions of licenses should be distinguished. First, licenses grant to other firms the legal right to use the technology patented by the licenser. Second, they may obligate the firm granting the license to provide technical information or in other ways assist the licensee in acquiring the technology. In return, the licenser normally receives royalties plus access to the innovations of the licensee.

In some European industries, established firms have used patents to block the entry of new firms. While this apparently has occurred in the receiving tube industry,[10] many of whose members are also important semiconductor producers, there is no evidence that patents have been so used in the European semiconductor industry. For a patent pool to succeed in limiting entry, the participation of Western Electric with its many strategic patents would be essential. This firm, showing no disposition to curtail entry, has licensed a multitude of firms in the United States and abroad.[11]

10. See Organisation for Economic Co-operation and Development, *Electronic Components: Gaps in Technology* (Paris: OECD, 1968), pp. 95–97.

11. Between 1952 and 1964, Western Electric licensed the Bell transistor patents to about 100 firms, of which the majority were foreign. See Great Britain, High Court of Justice, Chancery Division, "In the Matter of the Patents Act, 1949, and In the Matter of Letters Patent granted to Western Electric Company Incorporated . . . numbered 694,021 for Apparatus Employing Bodies of Semiconducting Material" (1964 W. No. 04; processed), Exhibits B, C, and D. For a list of the licensing agreements between American and British semiconductor firms with a

The Western Electric licensing policy is basically the same for foreign firms as for American firms. Agreements are negotiated on a firm-by-firm basis. Royalty rates and other conditions vary depending on the patents that the licensee can offer in exchange as well as on the potential of its R&D program. Firms like Texas Instruments and Philips, with important patents to trade in the semiconductor and other areas, pay reduced royalties or none at all, but even firms with nothing to exchange can obtain licenses and the royalties are not prohibitive. As with American firms, 2 percent on semiconductor sales is generally the maximum.[12]

The last chapter attributed the lenient Western Electric licensing policy to concern over antitrust, the desire to avoid patent infringement and litigation, and the belief that wide dissemination would accelerate progress in semiconductor technology. Also relevant to the licensing of foreign firms is the fact that AT&T confines its business operations to the United States and so has no incentive to restrict entry in foreign markets.

Other American firms, following the Western Electric licensing policy, have not tried to prevent the entry of new firms in Europe.[13] Fairchild, for example, is quite willing to license its basic planar patents. However, European firms have fewer Fairchild licenses than Western Electric licenses, partly because Fairchild asks for royalties as high as 4–6 percent of sales in addition to a lump-sum payment, and partly because many firms are willing to contest a patent infringement suit in court. Some firms even set aside a portion of sales in a contingency fund for possible litigation.

Texas Instruments, though it has licensed ITT and Philips, is in general more reluctant than Western Electric or Fairchild to license its semiconductor patents, including its basic integrated-circuit patents, to firms other than its own subsidiaries. Still, this has not kept Marconi Elliott

---

description of their terms, see Golding, "Semiconductor Industry in Britain and the United States."

12. Western Electric, which holds basic transistor circuit patents, has also charged European equipment firms royalties for using transistors in electronic products. American firms can obtain royalty-free licenses for these patents under the terms of the 1956 consent decree.

13. However, some American firms have tried to exclude foreign firms from the American market by prohibiting them from exporting to the United States semiconductors manufactured under licenses. For example, Fairchild's original licensing agreement with Elliott-Automation prohibited the English firm from exporting semiconductors to the American and Japanese markets. See "Elliott-Automation to Make Fairchild Planar Semiconductors," *Electronic Components*, Vol. 5 (November 1964), p. 984.

Microelectronics, Ferranti, Plessey, SGS, Sescosem, Telefunken, Siemens, and others from producing integrated circuits. Some of these firms, such as Marconi Elliott Microelectronics and SGS, have Fairchild licenses and are protected by the Fairchild–Texas Instruments accord not to dispute each other's patents. Others plan to fight in court if sued.

Although new firms have not been barred by patents from the semiconductor industry in Europe, licensing policies do favor the new foreign subsidiaries in several respects. With important patents to trade, Texas Instruments and Fairchild can obtain licenses at reduced or no royalties and can pass these savings on to their subsidiaries. In addition, American companies are likely to make a greater effort to assure the swift and complete transfer of technology to their foreign subsidiaries than to independent firms with which they have licensing agreements.

This brings us to the second function that licenses may perform, namely, the provision of technical assistance. As a general policy, Texas Instruments encourages the international transfer of its technology through its network of foreign subsidiaries. It believes that in the semiconductor field, where new technology may be obsolete in a year or two, this strategy maximizes the profits from new technological developments. Fairchild is more willing to provide technical assistance to independent firms, for royalties can be spent on research to help the firm maintain its technical leadership. Nonetheless, it is only natural that Fairchild should be more concerned with the transfer of know-how to its own subsidiaries than to other licensees.

On the other hand, AT&T has no European subsidiaries and hence no incentive to extend greater technical assistance to foreign subsidiaries than to European firms. In practice, it appears to promote the dissemination of its semiconductor technology as much to European firms as to American firms and their subsidiaries. As noted in Chapter 4, the firm has promptly announced its major breakthroughs and encouraged its professionals to contribute articles and books on semiconductor technology. These activities help firms on both sides of the Atlantic. In addition, Bell Laboratories invites representatives of foreign firms as well as American firms to visit its facilities. Finally, foreign firms (willing to pay the $25,000 in advance royalties) were invited to the 1952 and 1956 Bell symposia, which divulged much of the firm's semiconductor technology, including diffusion, oxide masking, and other important techniques. Among the participants at one or both of these sessions were AEI, Marconi, GEC, Ferranti, ITT, Pye, Philips, CSF, and Siemens.

Mobility of Scientists and Engineers

The mobility of scientists, engineers, and managers in the European semiconductor industry has increased in recent years particularly among the younger generation. Even so, mobility is still appreciably less than in the United States.

The lower incidence of scientists and engineers leaving established companies to form new firms is evident from the paucity of spin-offs in Europe, though there have been a few. Four employees of the Texas Instruments subsidiary in England left in 1961 to start Microwave Semiconductor Devices with financial backing from Microwave Associates in the United States. The effort collapsed the following year when the parent firm withdrew its support. Semitron, another British firm, was established in 1963. One of the three founders came from the Plessey research laboratory, another from Brush Crystal. In Germany, Eberle and Semikron entered semiconductor production around 1960 with the help of former employees from Intermetall and Siemens, respectively. Yet despite such exceptions, spin-offs in the European semiconductor industries are rare.

Mobility among established firms is also less in Europe. Companies are much more reluctant to recruit employees from rival firms. Even American subsidiaries do not appear as aggressive in this respect as firms in the United States. They hire employees from other firms when setting up their European facilities but, once established, tend to rely mainly on internal training and promotion to acquire indigenous specialists. This avoids alienating other semiconductor firms, many of which as large producers of final electronic equipment are potential customers.

Many factors contribute to the lower mobility in Europe. Certainly one is the attitude of European managers that recruiting the competition's personnel is unethical business behavior. Another is the paternalistic attitude of many European firms toward their employees. As a result, fringe benefits, such as vacation time, and even salary depend more on the length of service with the firm than is the case in the United States.

There are also institutional barriers to mobility. The small number of important semiconductor firms in any European country means the opportunities for changing jobs are limited. Many scientists and engineers, if they are going to leave their home country and subject their families to the language, educational, and other problems this involves, opt for the

United States, where salaries are higher and professional opportunities greater. In contrast, institutional factors forced some mobility on the American industry. Bell Laboratories, particularly in the early years, had by far the greatest number of scientists and engineers working in the semiconductor field, but the firm's growth in this area was stunted by the 1956 consent decree prohibiting sales in commercial markets. This limited the opportunities for Bell personnel and stimulated the large exodus from Bell Laboratories noted in the previous chapter.

Other possible reasons for the lower mobility in Europe are easy to find. Managers of European semiconductor firms frequently cite cultural differences and related sociological considerations. The educational system, particularly the training of engineers, may play a role. To identify all the factors and assess their relative impact on mobility would be a comprehensive study in itself.

The lower mobility of technical personnel within Europe and the emigration of many enterprising scientists and engineers to the United States have inhibited the rise of new firms in Europe, but not of new foreign subsidiaries. The latter have acquired the latest technology and know-how from American specialists provided by parent firms.

Thus, the mobility of scientists and engineers, like licensing policies, tends to favor new foreign subsidiaries. But the importance of both mobility and licenses can easily be exaggerated. For the low incidence of spin-offs, the migration of European scientists and engineers to the United States, and the superior know-how of American firms that makes licensing agreements with them and the services of their technical personnel desirable are all largely due to the different opportunities available to firms in Europe and the United States. The next section shows that in large part these differences arise because the nature and size of semiconductor markets vary in each of these areas.

Learning and Scale Economies, Market Characteristics, and Capital Availability

In the American semiconductor industry, economies of scale, though increasing in recent years, have not been large enough to bar the entry of small new firms. In contrast, learning economies have always been substantial, though their benefits are lost whenever new developments make old technology obsolete. The rapid succession of major innovations—the alloy process, silicon transistors, diffusion techniques, planar techniques, integrated circuits, epitaxial process, MOS techniques, and others—has provided new firms with frequent opportunities for entering the industry. However, the existence of learning economies has forced the new firms, at least the successful ones, to concentrate initially on new semiconductor devices and techniques where the cumulative production experience of the established firms is negligible.

In Europe, learning and scale economies in semiconductor production are similar in character and magnitude to those found in the United States. Yet the effect on entry of these economies, particularly learning economies, is not the same as in the United States because of the disparities in the size and nature of the semiconductor market and the availability of venture capital.

Market Size and Composition

The American and European semiconductor markets differ in several important respects. The first, as Figure 3-5 illustrates, is size. Semiconductor sales in Britain, France, and Germany, though growing in importance, are still only about one-tenth the value of American sales. Even combined, the European market is only about a third the size of the American market.

A second difference is the composition of the market. The proportion of semiconductor output destined for military equipment is much smaller in Europe.[14] Military equipment constitutes over half the American elec-

14. This is generally recognized and accepted in the industry, although data are not available for European countries. Even reliable breakdowns of final electronic markets by military, industrial, and consumer products which are comparable among countries are difficult to obtain. The figures cited below are based on market estimates by Texas Instruments found in Stewart Carrell, "International Report on Microelectronics," in Morton E. Goldberg (ed.), *Impact of Microelectronics –II,* Proceedings of the Second Conference on the Impact of Microelectronics (Electronic Industries Association and IIT Research Institute, 1968), Fig. 14. The data for the American market are consistent with those found in the *Electronic Industries Yearbook* (Washington: Electronic Industries Association, annual). This last source reveals that products for government (primarily military use) have accounted for substantially more than half of the American electronic market in the past. Data for the British market for the 1954–68 period are found in National Economic Development Office, Electronics Economic Development Committee, *Statistics of the Electronics Industry* (London: NEDO, June 1967), p. 95; and National Economic Development Office, *Annual Statistical Survey of the Electronics Industry* (London: NEDO, July 1969), p. 2.

tronics market, but less than a fifth of the European market. Though the share varies among the European countries, even in Britain and France, where it is relatively high, it does not approach half the total market. In Germany, the military market is small even by European standards.

Differences in market size and composition coupled with the typical shifting market pattern followed by most semiconductor devices produce a third important difference between the American and European markets. Many new semiconductor devices, like the silicon transistor and integrated circuit, are first used in significant quantities in military equip-

Figure 5-1. Value of Integrated Circuit Consumption as a Percentage of Semiconductor Consumption

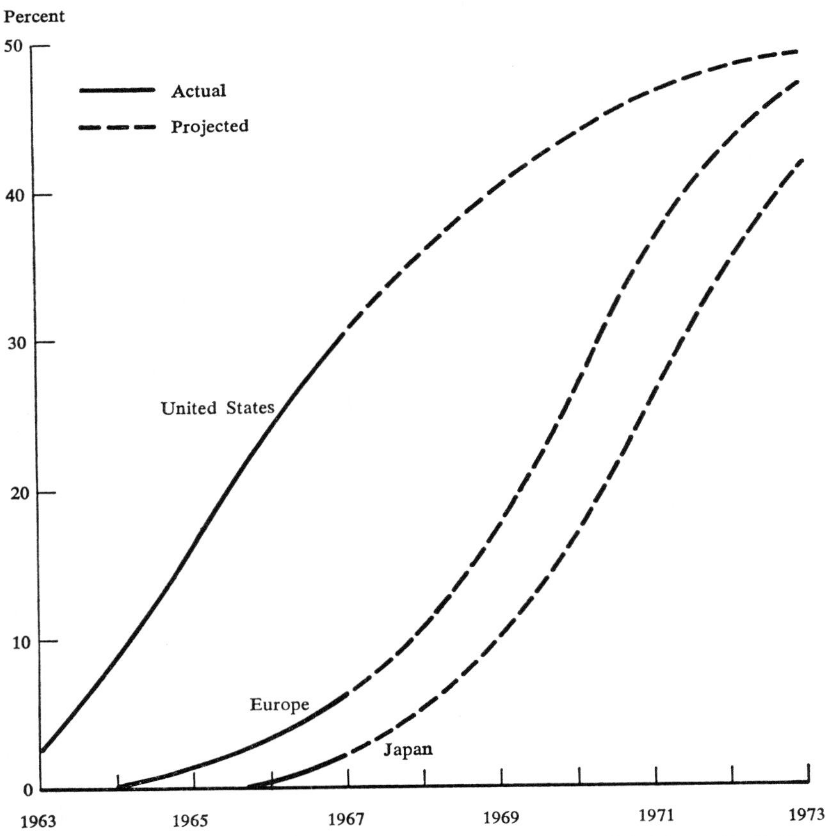

Source: Stewart Carrell, "International Report on Microelectronics," in Morton E. Goldberg (ed.), *Impact of Microelectronics–II*, Proceedings of the Second Conference on the Impact of Microelectronics (Electronic Industries Association and IIT Research Institute, 1968), Fig. 8.

ment. In time, as learning economies push prices down, demand arises in the industrial and consumer sectors of the market. Since the military market is larger in both relative and absolute terms in the United States, new devices generally penetrate the American market earlier and more extensively than the European market. Later, as demand arises from the industrial and consumer sectors, penetration proceeds faster in Europe (where the industrial and consumer markets are relatively more important) than in the United States, and the gap between the two closes. This pattern of lagging and then catching up, which characterizes Japan as well as Europe, is illustrated for integrated circuits in Figure 5-1.

As a result of these three differences, learning economies constitute a major barrier to the entry of new firms in Europe. Transitron, Texas Instruments, Fairchild, Signetics, and other new firms managed to enter the American market by concentrating on new devices, initially needed by the military. In Europe, this avenue for circumventing the competitive advantage learning economies give the established firms is closed. By the time significant demand arises in Europe for a new device, American firms with production experience can fill the European requirement at prices well below the costs of European firms.

The plight of the new European firm is aggravated by the division of the European semiconductor market into its national components and the preferential treatment some European governments bestow on their own producers. This fractionalizes the already small European government market and reduces even further the opportunities for new firms.

Trade with the United States does not provide an answer to this dilemma. Although transport costs and tariffs would not preclude new European firms from developing new devices and selling them in the American military market, the "Buy American" policy and other preferences the United States government extends to domestic producers do.[15]

Similarly, the opportunities for new European firms to exploit cost-reducing rather than quality-improving innovations have, at least so far, been limited. The demand for such innovations is substantial in Europe, where the cost-conscious commercial market is relatively important. However, most semiconductor innovations have the potential to both im-

15. See Chapter 3, note 14. European firms could circumvent these trade barriers by establishing or acquiring subsidiaries in the American semiconductor market. Despite the difficulties involved, Plessey apparently plans to try this approach. See Ted Schoeters, "A Widening Potential for Electronics and Automation," *Financial Times,* July 27, 1970.

prove quality and reduce costs, though the latter is usually realized only as production experience accumulates. The few significant exceptions, such as plastic encapsulation, are often labor-saving innovations, and strong economic forces encourage their introduction first in the United States with its high wage rate.

In short, the opportunities for European firms—new or established—to pioneer major new technologies and devices in the market are rare. Demand conditions greatly favor American firms in such endeavors.[16] Moreover, if European firms are to successfully imitate the American pioneers and thereby produce new semiconductor devices in Europe, large firms are needed, for such firms can absorb the initial losses often incurred in competing with American imports (see Figure 3-4), pay for the patents and technical assistance needed, and conduct R&D on a scale large enough to compensate at least in part for the lack of production experience.[17]

The one type of new firm that can surmount these obstacles is the new foreign subsidiary. In relation to American firms, these companies face less of a cost disadvantage than other companies in Europe because they can acquire from their American parents a greater portion of the know-how associated with learning economies. Moreover, their filial ties with American firms assure them of considerable technical assistance, R&D

16. The successful development of an innovation depends on supply as well as demand conditions. No matter how great the demand for an innovation, if it is beyond the state of the art, it will not be developed. Conversely, without demand, there is no incentive to produce an innovation, even though it may be relatively easy to do technically. Although both sets of forces affect the pace at which innovations are generated in an industry, economists differ over their relative importance. See Jacob Schmookler, *Invention and Economic Growth* (Harvard University Press, 1966); and Richard R. Nelson, "The Economics of Invention: A Survey of the Literature," *Journal of Business,* Vol. 32 (April 1959), pp. 101–27.

However, for the analysis here, it is not the overall rate of semiconductor innovation that is of interest, but rather the relative contributions of different countries to this rate. Because of the rapid and widespread dissemination of technical research through scientific journals and other media, the level of science is roughly the same in all industrially advanced countries. Consequently, demand conditions are primarily responsible for the differences among countries in introducing and diffusing new semiconductor technology.

17. An implicit assumption here is that large established firms can raise the capital needed for such high-risk, late-payoff projects, while small new firms, if they can obtain the necessary capital at all, can do so only on unfavorable terms. The next section indicates that venture capital has generally shunned the European semiconductor industry, depriving both large and small firms of external sources of funds. As an alternative, however, large firms can rely on internal financing.

support, and financial aid if necessary to weather a few lean years in the beginning.

Availability of Venture Capital

Concern that venture capital is inadequate, particularly for small and medium-sized firms, has troubled many European countries,[18] and a number of government agencies and other organizations have been created to alleviate any shortage. In the semiconductor industry, risk capital has not been as readily and extensively available in Europe as in the United States. Though the overall insufficiency of venture capital may be partly responsible, it apparently is not the whole explanation, since so little of the available venture capital has been invested in the semiconductor industry.

To attract capital, a risky business must offer some prospects for sizable profits and capital appreciation. In the United States, the success of Transitron, Texas Instruments, Fairchild, and others enticed many firms and persons to provide financial backing for new semiconductor efforts. In Europe, few semiconductor companies have been profitable, let alone outstandingly successful. Unlike American companies, they cannot amass large profits by introducing new semiconductor devices and reaping a quasi-monopoly rent until other firms are able to imitate. By the time demand in Europe is sufficient to support domestic production, European firms must compete with American firms and their subsidiaries, whose lower production costs force the European companies to accept low or even negative profits initially.

In addition to discouraging risk capital, low profits also deny European firms an important internal source of capital that would alleviate the need for risk capital. Successful American firms have reinvested a major portion of profits in their semiconductor operations.

Thus, the availability of venture capital has hindered the entry of new firms into the European semiconductor industry, but it, like the low mobility of scientists and engineers, is largely a manifestation of a more fundamental problem—the lack of market opportunities for new European firms. Profitable American subsidiaries have few problems obtaining capital from European sources. Though understandable, it is ironic that

18. See, for example, Organisation for Economic Co-operation and Development, Committee for Invisible Transactions, *Capital Markets Study: General Report* (Paris: OECD, 1967).

American subsidiaries should find it easier to raise European capital than European firms.

The Role of Government

The last section identified one important role European governments have not played—they have not provided the demand for semiconductors, particularly for new and sophisticated devices, that the American government has. This, along with the smaller size of the market in general, has hindered the entry of new European firms.

Despite the relatively low level of their own requirements, some European governments, especially Britain and France, have been greatly concerned about the fate of their semiconductor industries. Their interest dates back to World War II, when diodes were first used for radar equipment, but it has greatly increased in recent years with the extensive penetration of their markets by American subsidiaries. The realization that an independent computer industry, a high priority in both Britain and France, is not feasible as long as computer manufacturers must depend on American firms for integrated circuits and other essential components has added to this concern.

The British and French governments have tried to strengthen their semiconductor industries in a number of ways. This section focuses on two of the most important means—R&D support and industry rationalization—and their impact on entry conditions. The investigation is confined to Britain and France, for in Germany the government has shown much less concern over its semiconductor industry. There is no official rationalization policy (nor any need for one, since the German industry is already highly consolidated), and government R&D support is negligible.

R&D Support

The British and French governments support semiconductor R&D in two ways. First, like the American government, they fund a portion of the R&D work undertaken by private firms. Second, they maintain and operate their own laboratories. The latter, as Table 5-9 reveals, received in both countries in 1968 a larger share of the public funds for semiconductor R&D than companies did.

This table also shows that government funds are not trivial. In 1968,

Table 5-9. Semiconductor Research and Development Expenditures for Great Britain and France, 1968

Source of expenditure	Great Britain: Millions of dollars[a]	Great Britain: Percent	France: Millions of dollars[b]	France: Percent
Firm, self-financed	6.4	*43*	9.9	*55*
Firm, government financed	3.5	*23*	3.6	*20*
Government laboratories[c]	4.1	*27*	4.5	*25*
Universities	1.0	*7*	d	d
Total	15.0	*100*	18.0	*100*

Sources: British data are from Golding, "The Semiconductor Industry in Britain and the United States." The French figures are estimates based on information obtained from Bureau d'Informations et de Prévisions Economiques and the Ministère de l'Industrie, Paris.

a. Converted at the rate of $2.40 to the pound.

b. Converted at the rate of 4.9 francs to the dollar.

c. The two major British laboratories conducting semiconductor R&D are the Royal Radar Establishment under the Ministry of Technology and the Services Electronic Research Laboratory under the Admiralty. The major French laboratories are the Commissariat à l'Energie Atomique and the Centre National d'Etudes de Télécommunications.

d. Funds for semiconductor research conducted at French universities are included in the figures for government laboratories.

they totaled about $8 million each in Britain and France. This amount may equal more than a third of the semiconductor R&D funded by the American government.[19] Moreover, the proportion is even larger if the higher costs of conducting R&D in the United States are taken into account.[20] Also of interest is the relatively large share of all semiconductor R&D that the government sponsors in these two countries; in Britain, the government finances more R&D than the firms do.

One reason frequently cited for the rapid advance of semiconductor technology in the United States and the relatively strong position of

19. Comparable data for the United States are not available because an undetermined portion of government R&D support is passed on indirectly to semiconductor firms through prime contractors. Some fragmentary evidence, however, is available for recent years. See OECD, *Electronic Components: Gaps in Technology*, pp. 57–58.

20. R&D costs are probably between 50 and 100 percent higher in the United States than in Britain and France. See Christopher Freeman and A. Young, *The Research and Development Effort in Western Europe, North America and the Soviet Union* (Paris: OECD, 1965), Chap. 4 and App. 1.

Freeman and Young also found that R&D costs in 1962 were higher in France than in England. This was probably true for 1968 as well. The British had just devalued the pound. The French, however, did not devalue until 1969 so the franc was probably overvalued during 1968. As a result, the British may have devoted almost as much to semiconductor R&D in terms of real resources as the French, even though the latter spent 20 percent more in terms of dollars converted at the 1968 official exchange rate.

American firms throughout the world is the large amount of R&D supported by the American government. This may have been true in the past, but appears to be no longer, now that the British and French governments fund a substantial amount of semiconductor R&D.

Yet, for several reasons, government-sponsored R&D apparently is less productive, or at least has less commercial payoff, in these two European countries than in the United States. First, about half of the available public funds are absorbed by government and university laboratories, where commercial considerations are not as important in shaping the focus and direction of R&D activity as in company laboratories. Many projects carried out at the government and university laboratories involve basic research. Findings are generally published, which facilitates their quick dissemination throughout the world. This research becomes an input into the R&D efforts of foreign as well as domestic firms. (Indeed, American firms pushing on the frontiers of semiconductor technology may well benefit the most.) When public laboratories do produce new developments with direct commercial applications, problems of organization and communication may retard their transfer to firms for introduction into the market. To be sure, similar problems may plague large companies with central research laboratories not closely associated with production divisions. But even in such companies the problems are probably less imposing and the incentives to tap R&D results for commercial purposes greater.

Second, R&D contracts awarded to firms have emphasized research more and development less than have American R&D contracts. Even more important, unlike the production-refinement contracts in the United States, until recently government funds could not be used to help firms set up production lines. This policy has now changed, largely because of the realization that efficient production, as well as technical capability, is essential for a strong domestic semiconductor industry. Two contracts awarded around the end of 1967 signaled the change in Britain. One, to Elliott-Automation, covered the fabrication of MOS transistors; the other, to Ferranti, was for raising production yields.

Third, the European countries do not have the military demand for new and better semiconductors so prevalent in the American market. As a result, less R&D has been directed at hard requirements.[21] Integrated

21. The absence of hard requirements, of course, is partly responsible for the lack of emphasis on development and production projects in the allocation of past government R&D support.

circuits, for example, developed much more rapidly during the early 1960s in the United States than in Europe, largely because the United States was pushing its missile program, the first mass market for integrated circuits. American firms therefore had great incentives for striving to meet specific military requirements.

Some of the factors reducing the commercial payoff from government-sponsored R&D in Britain and France have also affected entry conditions. The large portion of R&D funds going to government laboratories reduces the support that could be given to new firms. Moreover, government laboratories have not been a source of trained scientists and engineers for new firms because the security and prestige associated with a civil service position keep mobility low.

Government R&D contracts favor the established firms even more than in the United States. Nearly all the British funds go to the GEC-English Electric combine (which includes AEI Semiconductors and Marconi Elliott Microelectronics), Mullard, Ferranti, Plessey, and STC (including Standard Telecommunication Laboratories). In France, the major recipients of government support are Sescosem, the CSF Central Laboratory (which is affiliated with Sescosem), and Silec.

Industry Rationalization

The American government has deliberately sought to prevent domination of the semiconductor industry by one or just a few firms. The Justice Department, for example, has forced AT&T to license any interested firm at reasonable royalties. In Britain and France, public policy has not only condoned consolidation, but in recent years even encouraged it. Marconi Microelectronics and Elliott Automation Microelectronics merged in 1968. The same year the new firm, Marconi Elliott Microelectronics, became an affiliate of AEI Semiconductors when its parent, English Electric, merged with the General Electric Company, which controls AEI Semiconductors. In France, the government pressured the General Electric Company of the United States (which is not associated with the British firm of the same name) to relinquish its 49 percent equity in Sesco, and then effected the merger of Sesco and Cosem in 1968. The government had hoped that Silec would also participate in the amalgamation, but so far the company has resisted.

To stimulate mergers and encourage collaboration in semiconductor R&D and manufacturing, the British and French governments are concentrating their R&D funds on a few large firms. The first awards under these new programs have already been made. The Ministry of Technology announced in 1969 that Marconi Elliott Microelectronics, Ferranti, and Plessey would receive, in roughly equal portions, $12 million over the next three and a half years. The funds are being invested through the National Research Development Corporation (NRDC), a public organization created to further technical developments in the national interest. Although NRDC will not hold stock in these companies, the amount of repayment depends on how successful the firms are. The $12 million is for R&D and production refinement projects in the manufacture of integrated circuits.

Similarly, the French government, pursuing its *Plan Composant,* promised Sescosem $18 million following its formation in 1968. The funds, to be allocated over five years, are specifically to help the company develop and produce integrated circuits for computer applications.

The aim of rationalization is to achieve at least one strong semiconductor firm in Britain and one in France that are independent of foreign interests and control. Public officials believe that only a large firm can fulfill this objective, and the last section supports this view. But there are several possible drawbacks to the present rationalization programs. First, they reduce the number of firms competing in the market. However, the remaining European firms are unlikely to stagnate from lack of competition, with the rising penetration of American subsidiaries into the European market. Second and more serious, combining a number of small, average firms may at best produce only a large average firm. The policy followed in the United States, which fosters the survival and growth of the fittest, is more likely to produce a large, superior firm. Yet in Europe this policy might leave only American subsidiaries operating in the semiconductor market, or at least in the more sophisticated sectors.

Third, the consolidation policy, which includes the concentration of government R&D support on a few established firms, impedes entry. Although new firms cannot compete in the production of semiconductor devices whose output has already reached substantial levels in the United States, they may be necessary if Europe ever hopes to take the initiative in diffusing or exploiting commercially new semiconductor devices. At least the important role played by such firms in the United States suggests this is the case. In the past, the small size of the European military market has limited the opportunities to pioneer the use of new devices. But this may change as semiconductor technology matures and becomes increasingly specialized. Possibilities for new devices designed specifically for industrial or consumer equipment may increase, enlarging Europe's opportunities to pioneer.[22]

In short, Europe faces a dilemma. Large firms are necessary to imitate American technology and compete with American subsidiaries. Without successful imitation there may be no European semiconductor firms. Yet if Europe wants eventually to pioneer the diffusion of new semiconductor devices, new firms and low entry barriers may be needed.

Conclusions

Three types of semiconductor firms can be identified in England, France, and Germany. The receiving tube firms, like their counterparts in the United States, are large, diversified companies with vested interests in first-generation active components. Nevertheless, they were among the first semiconductor producers in Europe. The new European firms are far less numerous than new firms in the American industry. Most entered the semiconductor field in the 1950s. Since then, new entries have been rare. Entry barriers, while rising for new European firms, have been falling for new foreign subsidiaries. There were few new subsidiaries at the beginning of the 1960s, but by the end of the decade their number had increased greatly.

Although the receiving tube firms have seldom introduced major semiconductor innovations, they have committed considerable resources to semiconductor R&D and obtained more semiconductor patents than the other two types of firms combined. During the 1950s, these firms were also the European leaders in the diffusion process. They supplied the bulk of the European requirements for semiconductor devices. During the 1960s, they were late in using silicon and planar techniques and lost a sizable segment of the market to new foreign subsidiaries. By swiftly penetrating neglected markets, the foreign subsidiaries assured the rapid

22. There is also some tentative evidence from other studies on the diffusion of new technology in the United States that suggests innovations spread more rapidly in less concentrated industries. See Edwin Mansfield, "Industrial Research and Development: Characteristics, Costs, and Diffusion of Results," in American Economic Association, *Papers and Proceedings of the Eighty-first Annual Meeting, 1968* (*American Economic Review*, Vol. 59, May 1969), pp. 69–70; and Edwin Mansfield, *Industrial Research and Technological Innovation: An Econometric Analysis* (Norton, 1968), Chap. 7.

diffusion of new semiconductor technologies in Europe. In the United States, new firms have performed this function, but in Europe, the new firms have generally remained small and specialized.

Among the factors impeding the entry of new European firms and facilitating the entry of new foreign subsidiaries is the greater access to state-of-the-art technology that the latter enjoy. Despite the liberal licensing policies followed by Western Electric and other American firms, foreign subsidiaries receive greater technical assistance and perhaps even reduced royalty charges because of their close ties with American parent companies. In addition, the low mobility of scientists and engineers within Europe hampers the ability of new firms to acquire the necessary technical personnel. New foreign subsidiaries, on the other hand, benefit greatly from the American specialists provided by parent firms.

The smaller market and limited amount of venture capital in Europe make economies of scale more significant than in the United States. But the principal barrier to entry for new European firms arises from learning economies. Unlike new American firms, new European firms cannot evade this barrier by concentrating on new devices. The pattern of demand in Europe tends to lag behind the pattern in the United States because the European military market is small. Consequently, American firms are the first to produce most semiconductor devices. By the time European demand justifies domestic production, American firms already enjoy considerable learning economies. Since American firms have greater access to this know-how, initially at least they have a competitive advantage over their European rivals.

Inadequate venture capital may increase the difficulty of starting a new European firm but seldom poses a problem for a new foreign subsidiary, which can usually rely on its parent for financing. The shortage of venture capital for new European firms is due primarily to the lack of market opportunities for these firms in the semiconductor field rather than to a general insufficiency of venture capital available in Europe.

Government R&D support and policies to rationalize the semiconductor industry also raise entry barriers in England and France. The largest share of public R&D funds goes to government laboratories in these countries, and consequently is not available to help launch new firms. Moreover, the funds granted to firms are concentrated among a few of the well-established domestic companies. This allocation reflects a conscious effort by the British and French governments to consolidate their semiconductor industries and create the large firms necessary to imitate

American technology and compete with foreign subsidiaries. Yet American experience suggests that easy entry conditions and new firms are often instrumental in pioneering the diffusion of new devices and techniques. Consequently, Europe may need easy entry conditions as well as large firms for a viable semiconductor industry which, in addition to responding to American initiatives, can also on occasion take the lead in diffusing new semiconductor developments.

CHAPTER SIX

Japan

THIS CHAPTER identifies the types of firms that have manufactured semiconductors in Japan and assesses their contributions to the innovative and diffusion processes. Unlike the chapters on the United States and Europe, it finds that the receiving tube firms are the principal diffusers of semiconductor technology in Japan. To uncover the reasons for this, the final sections of the chapter examine restrictions on direct foreign investment, other government policies, and the nature of the Japanese semiconductor market.

Types of Firms

Two major types of semiconductor firms can be distinguished in Japan, the receiving tube firms and new Japanese firms. There are no wholly owned foreign subsidiaries. Joint ventures with Japanese firms exist, but in no instance does the foreign partner hold a controlling interest.

Hitachi, Toshiba, Matsushita Electric, Nippon Electric, Mitsubishi Electric, and Kobe Kogyo (now part of Fujitsu) are the important producers of receiving tubes in Japan,[1] and all are active in the semiconductor industry. Except for Kobe Kogyo, which was only founded in 1949, these companies manufacture a wide range of products and rank among the largest electrical firms in Japan.[2] Like the receiving tube producers of

1. There are a few minor manufacturers, which in most cases are subcontractors to the major producers. See *1965 Japan Electronics Buyers' Guide* (Tokyo: Dempa Publications), Pt. 2, p. 1.

2. These companies are, however, somewhat smaller than the giants of the American electrical industry. Total 1968 sales for Hitachi, the largest of the Japanese firms, were surpassed by General Electric, IBM, ITT, Western Electric, Westinghouse, and RCA. See "The Fortune Directory of the 500 Largest Industrial Corporations," *Fortune* (May 15, 1969), pp. 166–87; and "The Fortune Directory: The 200 Largest Industrials Outside the U.S.," *Fortune* (Aug. 15, 1969), pp. 106–11.

America and Europe, their survival and growth do not depend on the receiving tube. Still, they have some vested interests in receiving tube technology, which is not true for Japanese semiconductor firms in the new firm category.

New firms in Japan, as in the United States and Europe, come in several varieties. Fuji Electric was large and diversified like the receiving tube producers when it entered the semiconductor business. A few firms such as Kyodo were established to produce semiconductors. Sony, Sanyo, and most of the new firms were small or medium-sized concerns established in other industries at the time they moved into the semiconductor field. Sony, for example, was a successful small producer of tape recorders and magnetic tape; Sanyo was a medium-sized firm known for its washing machines; Fujitsu manufactured communications equipment; and Japan Radio produced special purpose tubes.

The six receiving tube producers and twelve new firms and the years they have engaged in the commercial production of semiconductor devices are identified in Table 6-1. Although all the important semiconductor manufacturers are included, the table probably omits some minor producers.[3] It is also possible that the production dates attributed to several firms are not precisely accurate. Despite these limitations, the table does illustrate several interesting features in the development of the Japanese semiconductor industry.

First, the six receiving tube producers were among the earliest firms in the country to undertake semiconductor production. By 1957, all were active in the industry. Though Sony, a new firm, was the first to produce commercial transistors in Japan, its initiative was quickly followed by the receiving tube firms.

Second, all but one of the new firms entered the industry in the second half of the 1950s and early 1960s. The table suggests that entry barriers have since become more formidable.

Third, entry barriers for foreign subsidiaries apparently are prohibitive. The table shows no wholly owned foreign subsidiaries. It identifies a number of joint ventures, such as Matsushita Electronics, New Japan Radio, and the International Rectifier Corporation of Japan, but in no instance does the foreign partner hold a controlling interest.

3. The *1965 Japan Electronics Buyers' Guide* lists sixteen firms as manufacturers of transistors, diodes, and rectifiers in addition to the firms shown in the table. None is an important producer. Some, though listed as manufacturers, may be wholesalers or importers. Others may be subsidiaries or affiliates of larger semiconductor firms.

Table 6-1. Semiconductor Firms in Japan and Years They Were Active in the Industry, 1954–68[a]

Firm	*1954*	*1955*	*1956*	*1957*	*1958*	*1959*	*1960*	*1961*	*1962*	*1963*	*1964*	*1965*	*1966*	*1967*	*1968*
Fujitsu[b]	x	x	x	x	x	x	x	x	x	x	x	x	x	x	x
Kobe Kogyo[c]	x	x	x	x	x	x	x	x	x	x	x	x	x	x	x
Nippon Electric[d]	x	x	x	x	x	x	x	x	x	x	x	x	x	x	x
Sony[e]	x	x	x	x	x	x	x	x	x	x	x	x	x	x	x
Toshiba[f]	x	x	x	x	x	x	x	x	x	x	x	x	x	x	x
Hitachi[g]	x	x	x	x	x	x	x	x	x	x	x	x	x	x	x
Origin Electric			x	x	x	x	x	x	x	x	x	x	x	x	x
Japan Radio[h]				x	x	x	x	x	x	x	x	x	x	x	x
Matsushita Electric[i]				x	x	x	x	x	x	x	x	x	x	x	x
Mitsubishi Electric[j]				x	x	x	x	x	x	x	x	x	x	x	x
Oki Electric						x	x	x	x	x	x	x	x	x	x
Sanyo[k]							x	x	x	x	x	x	x	x	x
Fuji Electric							x	x	x	x	x	x	x	x	x
Shindengen Electric								x	x	x	x	x	x	x	x
Yaou Electric								x	x	x	x	x	x	x	x
Sanken Electric									x	x	x	x	x	x	x
International Rectifier (Japan)[l]									x	x	x	x	x	x	x
Kyodo												x	x	x	x

Sources: *Japan Electronics Buyers' Guide* (Tokyo: Dempa Publications, annual); *Electronics in Japan* (Tokyo: Electronics Association of Japan, annual); other trade publications; correspondence with firms.

a. The specific years each firm has engaged in the commercial production of semiconductors are denoted by "x." Names of subsidiary companies that also produce semiconductors and former company names that have been changed are indicated in the following notes.

b. Fujitsu (full name, Fuji Tsushinki Seizo K.K.) began commercial production of diodes about 1953 and transistors soon thereafter. Large-scale transistor fabrication began in 1957.

c. The Kobe Kogyo Corporation was producing commercial transistors by 1954. The firm merged with Fujitsu in 1968.

d. Began large-scale production of diodes in 1953 and transistors in 1955. The New Nippon Electric Company, a subsidiary, also produces semiconductors. International Telephone and Telegraph (U.S.) holds a 12 percent interest in the parent firm.

e. Called Tokyo Tsushin Kogyo until about 1958; first firm to produce commercial transistors in Japan.

f. Toshiba (full name, Tokyo Shibaura Electric Company) had transistors on the market by 1954. Mass production began about 1956. General Electric (U.S.) has about a 10 percent interest in Toshiba.

g. Was producing commercial semiconductors by about 1954.

h. The New Japan Radio Company, a joint venture of the Japan Radio Company and Raytheon (U.S.), also produces semiconductors. The American firm owns a one-third interest.

i. Matsushita Electric Industrial Company controls Matsushita Electronics, another semiconductor manufacturer. Philips (Netherlands) owns a minority share (30 percent).

j. Mitsubishi Electric Corporation has close ties with Westinghouse (U.S.), which holds a small interest (around 4 percent).

k. The Tokyo Sanyo Electric Company, a subsidiary of the Sanyo Electric Company, also produces semiconductors.

l. International Rectifier (U.S.) owns 33 percent of the International Rectifier Corporation of Japan.

Fourth, the casualty rate for semiconductor firms is very low. The only firm that Table 6-1 identifies as no longer an independent semiconductor producer is Kobe Kogyo, which merged with Fujitsu in 1968.[4]

Contributions by Firms

This section examines the performance of Japanese semiconductor firms in the innovative and diffusion processes, and evaluates the relative contributions of the receiving tube firms and new firms to each of these activities.

Innovative Process

Semiconductor R&D expenditures of individual Japanese firms are not available. However, certain observations are frequently encountered in the trade press and elsewhere regarding Japanese R&D in the semiconductor field and are noted before we pass on to other measures of innovative performance where more data are available.

RESEARCH AND DEVELOPMENT. All the receiving tube firms except Matsushita and Mitsubishi had initiated semiconductor R&D programs by the early fifties,[5] and by 1953 they had been joined by Sony and Fujitsu. Despite the early interest in the semiconductor field of most of the receiving tube producers and a few new firms, the amount of R&D expenditures remained quite low throughout the fifties. Firms relied primarily on technical assistance agreements with foreign companies to acquire the necessary technology and devoted most of their limited R&D outlays to absorbing foreign know-how. Though R&D expenditures have since increased considerably, they still amounted to only 2 percent of semiconductor sales in the mid-sixties, a time when the United States was spending 6 percent or more.[6]

In allocating R&D resources, Japanese firms with few exceptions have stressed development and engineering projects. Modern central research

4. Even before the merger, Kobe Kogyo had ties with Fujitsu through their connection with the Dai Ichi Bank.

5. See C. C. Gee, "World Trends in Semiconductor Development and Production," *British Communications and Electronics,* Vol. 6 (June 1959), pp. 450–61.

6. See Organisation for Economic Co-operation and Development, *Electronic Components: Gaps in Technology* (Paris: OECD, 1968), Table B2; and note 10 in Chapter 4, above.

laboratories at Hitachi, Nippon Electric, Matsushita, Toshiba, and Mitsubishi Electric, with some of Japan's best technical talent, have concentrated largely on minor advances in proven technology, often with relatively short payoff periods. This emphasis has produced many ingenious additions to imported technology but few major advances.

This R&D strategy, which also prevails in other industries, has used Japan's limited R&D resources efficiently. But in recent years, the cost, restrictions, and difficulties encountered in obtaining foreign technology have increased. In addition, as Japanese technology has advanced rapidly, approaching the best technology found abroad in many fields, the pool of foreign know-how the country can draw on for further advances in its technology has dwindled. As a result, the government is now stressing the need for more research, and the traditional R&D strategy may be changing.

PATENTS. American semiconductor patents[7] awarded to Japanese producers are shown in Table 6-2. American patents are considered primarily for convenience,[8] though they have another advantage over Japanese patents as well. Like other firms, Japanese companies patent major inventions throughout the world, but many minor developments only in Japan. Fujitsu, for instance, has more than ten times as many Japanese patents as American patents in the semiconductor field.[9] Consequently, American patents should provide a better measure of interfirm differences in generating significant advances in technology.[10]

7. These patents cover new semiconductor devices, methods of production including materials preparation, and manufacturing and testing equipment, but exclude new applications of semiconductors in final electronic products.

8. Japanese semiconductor patents are found under a number of different classifications. This makes their collection time consuming and increases the possibilities for error.

9. This information was obtained through correspondence with the firm.

10. It is possible that the propensity to patent in the United States varies from one Japanese firm to another. If receiving tube firms are either more or less prone to apply for American patents than new firms, this would bias any conclusion based on Table 6-2 about company contributions to the innovative process. To check for this possibility, Japanese semiconductor patents were collected for 1962. Twenty-one semiconductor patents were found for Hitachi, fifteen for Nippon Electric, fourteen for Sony, thirteen for Japan Radio, six for Toshiba, four for Origin Electric, three for Oki Electric, two for Sanyo, two for Kobe Kogyo, two for Shindengen, and one for Fuji Electric. See the Japanese Patent Office's Patent Record for 1962 (in Japanese; translated for the author).

The Japanese patents for only two firms appear surprising when compared with their American patents around the same period. Hitachi, a receiving tube firm, and

Table 6-2. U.S. Semiconductor Patents Awarded to Firms in Japan, 1959–68[a]

Type and name of firm[b]	*1959*	*1960*	*1961*	*1962*	*1963*	*1964*	*1965*	*1966*	*1967*	*1968*	*Total, 1959–68*
Receiving tube firms											
Nippon Electric				3	3	7	11	11	8	6	49
Matsushita Electric									3	5	8
Hitachi								1	1	1	3
Toshiba				1				1			2
Subtotal				**4**	**3**	**7**	**11**	**13**	**12**	**12**	**62**
New firms											
Sony	1	3		3	3	1	1	2		1	15
Fujitsu								2	1	1	4
Sanyo						1	1				2
Subtotal	**1**	**3**		**3**	**3**	**2**	**2**	**4**	**1**	**2**	**21**
Total	**1**	**3**		**7**	**6**	**9**	**13**	**17**	**13**	**14**	**83**
Percentage of total											
Receiving tube firms	*0*	*0*		*57*	*50*	*78*	*85*	*76*	*92*	*86*	*75*
New firms	*100*	*100*		*43*	*50*	*22*	*15*	*24*	*8*	*14*	*25*

Source: U.S. Patent Office records.

a. The period 1954–58, included in Tables 4-2 and 5-4, is not included here since no patents were awarded to Japanese firms in those years. Nor were any patents awarded in 1961.

b. Firms identified in Table 6-1 that are not listed in this table received no American semiconductor patents during 1959–68.

The table indicates that the receiving tube firms as a group have generated nearly three times as many patents as the new firms.[11] However, the variance among firms in each of these groups is great. Nippon Electric, a leader in integrated-circuit technology in Japan, alone accounts for 79 percent of the patents granted to the receiving tube firms, and Sony for 71 percent of the patents awarded to the new firms. Conversely, there are firms in both groups with no American patents in the semiconductor field.

Japan Radio, a new firm, received more Japanese patents than might be expected on the basis of their American patents. Table 6-2, therefore, may underestimate the contributions of these firms to the innovative process.

11. This does not necessarily mean that the innovative performance of the Japanese semiconductor industries would be greater if there were more or larger receiving tube firms and fewer new firms. Before such a judgment is made, the number of patents that firms have should be adjusted for the differences in their market shares identified below. Factors such as the effect of entry conditions and new firms on the innovative activity of receiving tube firms should also be considered.

The table also reveals that few patents were awarded before 1962, and none before 1959. Even after adjusting for the patent pendency period of three to four years, this supports the observation made above that Japanese semiconductor R&D was quite low throughout the fifties and devoted primarily to assimilating foreign technology.

The first Japanese producer to acquire semiconductor patents was Sony. During the sixties, however, Nippon Electric dominated the acquisition of patents, with Matsushita becoming increasingly important toward the end of the decade. Consequently, unlike the patent scenario in the United States and Europe, the receiving tube firms in Japan, thanks to a few of their members, have over time increased the proportion of new semiconductor patents they have received, while the share of the new firms has fallen.

MAJOR INNOVATIONS. The only major innovation, as identified in Table 2-1, introduced by a Japanese firm is the tunnel diode developed at Sony in 1957[12] This finding is consistent with two earlier observations on Japanese innovative activity in the semiconductor field. First, Sony appears to have made the greatest contribution to the innovative process during the 1950s. Second, the Japanese have concentrated their R&D efforts on absorbing foreign technology and making modest improvements on it.

Diffusion Process

Expenditures for R&D, patents, and major innovations all reflect the heavy dependence of Japanese producers on imported technology, and in turn the importance of diffusion for the Japanese industry. This section examines the performance of firms in the diffusion process, considering first the intercountry transfer of new technology and then its intracountry diffusion.

Four firms—Sony, Fujitsu, Nippon Electric, and Toshiba—initiated the commercial production in Japan of the thirteen major semiconductor devices identified in Table 3-1, and thereby brought about the intercountry diffusion of these innovations. The two new firms, Sony and Fujitsu, were particularly instrumental in introducing devices developed in the 1950s. Sony, for example, pioneered the manufacture of transistors in

12. Actually, as note e, Table 3-1, indicates, although Sony invented the tunnel diode, it was first produced commercially in the United States about 1960 by RCA and General Electric.

Japan after the interest of its president in this new development was aroused during a visit to the United States in 1952. Conversely, the receiving tube producers have introduced most of the devices of the 1960s. The Japanese leader in integrated circuits, for instance, is Nippon Electric. The growing importance of the receiving tube producers in effecting the intercountry diffusion of semiconductor technology to Japan runs counter to the trend found in Europe.

Over the entire period, as Table 6-3 indicates, the Japanese receiving tube producers have initiated a greater proportion of the major semiconductor devices than the new firms. In addition, their average imitation lag has been about a third of a year shorter, but comparisons of imitation lags are risky. There is considerable variation within groups, and Sony's lag of just over one and a half years is appreciably shorter than that of either of the receiving tube firms. Moreover, if the surface-barrier transistor with its unusually long lag is dropped from the calculations, Fujitsu's lag shrinks to one and a half years, making the average lag for the two new firms nearly a full year shorter than the average lag of the receiving tube firms.

Because of the heavy reliance of Japanese semiconductor producers

Table 6-3. Percentage of Major Semiconductor Devices First Produced in Japan, by Firm, and Average Imitation Lags, 1953–68

Type and name of firm	*Percentage of innovations initiated*[a]	*Average imitation lag (years)*
Receiving tube firms		
Nippon Electric	46	2.25
Toshiba	12	3.00
Subtotal	**58**	
Subaverage		**2.40**
New firms[b]		
Sony	27	1.57
Fujitsu[b]	15	4.75
Subtotal	**42**	
Subaverage		**2.73**
Total and average	**100**	**2.54**

Source: Table 3-1.

a. When two firms are responsible for the first production of a major semiconductor device, each is given one-half the credit.

b. The imitation lag for the surface-barrier transistor, which Fujitsu produced first in Japan, was an unusually long eight years. If this lag is excluded, Fujitsu's average imitation lag is 1.50 years, and the average imitation lag for the new firms as a group is 1.55 years.

Table 6-4. Semiconductor Market Shares in Japan, by Firm, 1959 and 1968[a]

Type and name of firm	Percentage of market	
	1959	*1968*
Receiving tube firms		
Toshiba	26	21
Matsushita Electric	16	15
Hitachi	15	23
Nippon Electric	15	7
Kobe Kogyo[b]	5	3
Mitsubishi Electric	2	3
Subtotal	**79**	**72**
New firms		
Sony	11	2
Sanyo	2	13
Fujitsu[b]	1	1
Others	5	2
Subtotal	**19**	**18**
Importers	**2**	**10**
Total	**100**	**100**

Sources: The market share filled by imports is found in the sources noted for Figures 3-2 and 3-5. The 1968 market shares for Japanese firms are based on semiconductor production during December of that year, which is given in Survey of the Japanese Electronic Component Market, 1969 (Tokyo: Science Newspaper Company), p. 664 (in Japanese; translated for the author). The 1959 market shares are based on data found in Marketing Research Branch, Texas Instruments, Inc., "In Support of the Electronics Industries Association Petition before the Office of Civil and Defense Mobilization," prepared for the Imports Committee, Semiconductor Section, Electronics Industries Association, Supplemental Research Report (Texas Instruments, Dec. 15, 1959), pp. 33–36.

a. Market shares for Japanese firms (but not importers) are based on the number of semiconductor devices produced rather than the value of sales, because data on the latter are lacking. As a result, the market shares of firms concentrating on expensive, special semiconductors or devices for the industrial market are somewhat understated.

b. Although Kobe Kogyo and Fujitsu merged in 1968, their individual market shares for that year were estimated and are shown separately.

on foreign technology, the ability of individual firms to compete in the Japanese market depends greatly on their proficiency in using the important new developments from overseas. Consequently, market shares provide an approximate measure of company performance in the intracountry diffusion of semiconductor technology.

Market shares for the major firms in the Japanese industry are shown for 1959 and 1968 in Table 6-4. It indicates that the receiving tube producers have dominated the Japanese market, accounting for 79 percent of the market in 1959 and 72 percent in 1968. The share held by the new firms has fallen from 19 to 18 percent, while imports have grown from 2 to 10 percent.

The three largest producers in 1968 were Hitachi, Toshiba, and Matsushita—all receiving tube firms. They were followed by Sanyo, the only major producer from the new firms. Established about 1950, this fast-growing company greatly increased its share of the semiconductor market in the 1960s, while Sony, producing mostly for in-house needs, saw its share fall.

The receiving tube firms have captured the largest market shares by emphasizing the large-volume markets. Most of the new firms have concentrated on special devices with limited markets. Fujitsu, for example, did not produce semiconductors for the large consumer market until it absorbed Kobe Kogyo in 1968. Origin Electric has specialized in rectifiers, Oki Electric in industrial transistors, and Shindengen in power devices.

The success of the receiving tube firms in capturing and holding large market shares that correspond to or even exceed their relative contribution to the innovative process contrasts sharply with the situation in the United States and Europe. This, along with the important contributions the receiving tube firms have made in the intercountry transfer of new innovations, indicates that they have been largely responsible for the diffusion of semiconductor technology in Japan. The remainder of this chapter examines the principal factors responsible for this finding, beginning with the important role played by the Japanese government.

The Role of Government

Previous chapters have shown that government policies and programs affect the ability of the various types of firms to enter and prosper in the semiconductor industries of Europe and the United States. This is true for Japan as well. Indeed, the influence of the government permeates Japanese industry to an extent unknown in the Western countries.

Control over foreign trade and the importation of capital and technology give the government considerable power to intervene directly in the affairs of industry and individual enterprises. In addition, public officials can exert considerable indirect pressure through industry trade associations and through the many informal consultations that take place between the various government ministries and business firms.[13]

13. For more on the influence of the government in the electronics industry and industry generally, see Frank Leary, "Electronics in Japan," *Electronics*, Vol. 33 (May 27, 1960), pp. 53–100; William W. Lockwood, "Japan's 'New Capitalism,' "

Government and Foreign Subsidiaries

Under the foreign investment laws enacted in the early postwar period, the government approves applications for all direct investments by foreigners. In the semiconductor industry, it has turned down all requests for wholly owned subsidiaries and for joint ventures where the foreign firm retains over 50 percent of the equity.[14] Rarely, in fact, has the foreign partner been allowed to hold more than a one-third interest. Even joint ventures with minor foreign participation are seldom approved unless there are appreciable benefits, notably access to new technology, for Japan. Similarly, the government has restricted foreign purchases of stock in established Japanese firms to assure that their control does not fall into the hands of foreign interests.

After the war, the Japanese justified the restrictions on the inflow of foreign capital on the grounds that subsequent withdrawal of such capital would adversely affect their balance of payments. However, as the Japanese balance of payments has improved, it has become increasingly clear that a primary reason for curbing direct foreign investment is the fear that foreign interests will gain control over important segments of the Japanese economy. The belief that foreign technology can be acquired more cheaply through licenses has also encouraged restrictions on direct investments.

The government's determination to resist the formation of semiconductor firms controlled by foreign interests was tested in the late 1960s by Texas Instruments. The American firm petitioned the Japanese government for a wholly owned subsidiary in the early 1960s, and was offered instead a minority interest in a joint venture with a Japanese company. Texas Instruments rejected this offer and continued to press for a wholly owned subsidiary. To strengthen its bargaining position, it refused to license its strategic integrated-circuit patents to Japanese firms.[15] This

in Lockwood (ed.), *The State and Economic Enterprise in Japan: Essays in the Political Economy of Growth* (Princeton University Press, 1965), pp. 447–522; and Eleanor M. Hadley, *Antitrust in Japan* (Princeton University Press, 1970), Chap. 16.

14. This is generally true for other industries as well, though there are some exceptions. The IBM facility in Japan, established before World War II, is a wholly owned subsidiary.

15. Nippon Electric did have a Fairchild license and in turn had granted sublicenses to a number of other Japanese firms. This did not, however, protect the Japanese firms, as it did Fairchild licensees in Europe and the United States, against the possibility of a patent infringement suit by Texas Instruments, because the Fairchild-Texas Instruments patent accord explicitly excluded Japan.

did not prevent the latter from manufacturing integrated circuits, because the applications submitted by Texas Instruments for Japanese patents covering integrated circuits became enmeshed in administrative procedures and were delayed for many years. But when Japanese electronic firms began exporting electronic calculators and other equipment with integrated circuits in 1967, the Japanese government became concerned that the American firm might start court action in the United States, claiming its American patents had been violated, and so the government forced the Japanese firms to withhold their exports. By this time, though, several Japanese firms were actively engaged in integrated-circuit fabrication. Protected by government restrictions banning imports of many types of integrated circuits, these firms were steadily gaining greater proficiency, making the task of breaking into the Japanese market more difficult for Texas Instruments the longer it delayed.

Thus there were pressures on both sides to find a settlement, and in 1968 a compromise was reached, though most of the concessions were made by Texas Instruments. The firm dropped its demand for a wholly owned subsidiary and agreed instead to a joint venture with Sony, with each firm holding 50 percent of the equity. In addition, it promised to license its integrated-circuit patents to Nippon Electric, Hitachi, Mitsubishi, Toshiba, and Sony and to limit its production so that 90 percent or more of the Japanese integrated-circuit market would be left for these firms.[16] For its part, the Japanese government allowed Texas Instruments to retain a 50 percent interest in the new joint venture. No other foreign firm holds such a large share in a Japanese semiconductor firm. The closest are Philips with a 30 percent interest in Matsushita Electronics, Raytheon with a 33 percent interest in the New Japan Radio Company, and International Rectifier with a 39 percent interest in the International Rectifier Corporation of Japan.[17] Most other foreign interests are under 15 percent. For instance, ITT holds a 12 percent share of Nippon Electric, and General Electric a 10 percent share of Toshiba.

Government and Japanese Firms

The government affects the ability of domestic firms, as well as foreign firms, to enter and compete in the semiconductor industry. Its more im-

16. See "International Outlook," *Business Week* (May 4, 1968), p. 100; and "Around the World," *Electronics* (May 13, 1968), p. 200.

17. The share of equity held by these foreign firms as well as those noted next are found in *Moody's Industrial Manual* (Moody's Investors Service, 1970) or in OECD, *Electronic Components: Gaps in Technology*, p. 186.

portant activities in this regard are four: influencing companies' access to foreign technology through licenses, R&D activity in the semiconductor field, industry rationalization, and competition.

LICENSES. The government approves all licensing agreements between Japanese and foreign firms, which gives it considerable control over companies' access to foreign technology. It has authorized many such agreements in the semiconductor industry.[18] Foreign companies, for their part, have usually been willing to license Japanese firms and even to assist in the transfer process.

Nearly all the Japanese firms have had Western Electric licenses since they began semiconductor production. Many have also benefited from agreements with RCA and General Electric, and recently a number have acquired Fairchild and Texas Instruments licenses. While these are the principal firms to which the Japanese have turned for licenses in the semiconductor area, there are others. Mitsubishi Electric, for example, has an agreement with Westinghouse, and Matsushita with Philips.

Licenses grant Japanese firms the legal right to use the many fabrication processes and to produce the many devices that foreign firms have developed and patented, and enable them to enlist the assistance of foreign firms in effecting the actual transfer of technology by providing for detailed production information, specifications, and even manufacturing equipment. The government's willingness to approve licenses extends to both those covering the use of patents and those providing for technical assistance.[19] When necessary, it has allowed the Japanese firms to pay substantial royalties. For example, all integrated circuit producers pay 10 percent of sales—2 percent to Western Electric, 4.5 percent to Fairchild, and 3.5 percent to Texas Instruments.[20]

18. The license agreements approved by the government, the Japanese and foreign firms involved, and the technology covered are listed in Japanese Electronics Yearbook (Tokyo: Dempa Publications, annual; in Japanese; translated for the author).

19. The government, however, does not approve all license requests. It may, for instance, reject (or rather send back to the negotiating table) an application if it believes the royalty fee to be too high. One advantage of government control over licenses is the monopsonistic power the government can exercise in bargaining with foreign firms.

Agreements that discourage domestic R&D may also be rejected. As the government became increasingly concerned in the 1960s about the cost and restrictions associated with foreign technology, it began to encourage more domestic R&D and to scrutinize license applications more closely.

20. "Major Japanese Firms Win Better Deals from U.S. Ties," *Electronic News* (March 24, 1969), p. 32.

Whether the government has favored either the receiving tube firms or the new firms in authorizing licenses is difficult to determine. Both types of firms have numerous foreign licenses, and it seems unlikely that the large share of the semiconductor market enjoyed by the receiving tube firms can be attributed to government discrimination that denies new firms access to vital technology from abroad. In one respect, the government has even helped the less established new firms. As a general policy, designed to restrict the development of monopolies, it requires foreign firms to license all Japanese firms requesting (with the approval of the government) a license and to charge one royalty rate. In other countries, it will be recalled, royalty rates are negotiated on a firm-by-firm basis, and normally the established firms with important patents and large R&D operations obtain lower royalties than new firms with nothing to offer in exchange. In Japan, this is not the case. Western Electric, for example, charges all its Japanese licensees 2 percent of semiconductor sales.

R&D SUPPORT. The amount of government-funded R&D conducted by firms in the semiconductor field is trifling in Japan,[21] and so the distribution of these funds probably has little effect on firm competitiveness. This is even more likely in light of the government requirement, attached to at least some of its R&D contracts, that the recipient firm share its findings and results with competitors in order to minimize duplication of effort and conserve the country's R&D resources.[22]

Although it supports little R&D by companies, the government does finance the research activities of the state-run universities and a number of government laboratories, such as the Electrotechnical Laboratory and the Electrical Communications Laboratory. Much of the best semiconductor research in Japan, particularly fundamental research, has been carried out at these facilities. But again, to avoid duplication of R&D effort, research results have generally been available to all Japanese firms, and so it is unlikely that the R&D conducted at government facilities has enhanced the technological capabilities of the receiving tube firms much more than those of new firms.

INDUSTRY RATIONALIZATION. Conflicting pressures and concerns have shaped Japanese public policy in the area of industry rationalization since World War II. Some officials, particularly those in the Ministry of International Trade and Industry (MITI), worry about the inefficiency aris-

21. See OECD, *Electronic Components: Gaps in Technology,* p. 178.

22. See, for example, Yasuo Tarui, "Japan Seeks Its Own Route to Improved IC Techniques," *Electronics* (Dec. 13, 1965), pp. 90–91.

ing from fragmented market structures and small firms operating below optimum scale. Like their counterparts in Europe, they are urging industry consolidation.

Other officials, particularly those in the Fair Trade Commission, are concerned about maintaining and stimulating competition, and they oppose consolidation. To a great extent, this concern with competition is a legacy of the American occupation. The Japanese antimonopoly law was passed during this period to dissolve the large *zaibatsu,* or trusts, that dominated Japanese industry before World War II. These family-run holding companies had subsidiaries and affiliates throughout the economy.

Since 1952, when Japan regained its independence, MITI and its allies have managed to dilute the antimonopoly law on several occasions, and the zaibatsu have revived to some extent. The old ruling families and holding companies are gone, however, and today control is much more loosely exercised.[23]

The resurgence of the zaibatsu and the pressures from MITI and other sources for industry rationalization may have inhibited the entry and growth of new firms in the semiconductor industry, but certainly not greatly. Table 6-1 shows that new firms managed to enter the semiconductor industry throughout the fifties and sixties, that no firms were forced to abandon the industry, and that only one merger took place. These are not features that characterize an industry under heavy pressure to rationalize.

COMPETITION. The government has actively promoted competition among the existing semiconductor firms. Its refusal to allow domestic firms to establish monopoly positions based on technology created by government-sponsored R&D or acquired by license from foreign companies is part of this effort.

The government's success is reflected by the market shares shown in

23. It is commonly presumed that control is now exercised by the large banks affiliated with these groups. Their influence is assumed to arise primarily from their lending power rather than equity holdings, for Japanese firms rely much more heavily on debt financing than American or European firms.

Among semiconductor producers, Mitsubishi Electric is affiliated with the Mitsubishi bank, Nippon Electric with the Sumitomo bank, and Fuji Electric and Fujitsu (and Kobe Kogyo before it merged with Fujitsu) with the Dai Ichi bank. Toshiba, Hitachi, Matsushita, Sanyo, Sony, and others also depend on banks for capital. But rarely do Japanese firms rely on only one bank. Thus the power of banks to influence company policies by threatening to withhold funds is probably marginal. This has led some writers to challenge the view that banks are the new control centers of the zaibatsu. See Hadley, *Antitrust in Japan,* Chap. 11. For a different view, see Leary, "Electronics in Japan."

Table 6-4. No Japanese firm has captured close to half the market, as Philips did in Europe before the arrival of American subsidiaries. Instead, four or five firms have shared sizable and roughly equivalent portions of the market. Moreover, the size and ranking of market shares have changed over time as firms have battled with varying degrees of success to increase their sales. In the process, one new firm, Sanyo, has managed to acquire a significant share. This suggests that the government's effort to foster competition among Japanese firms has helped new firms compete against established producers after entering the industry.

Thus it is evident that the hand of the government is felt throughout the Japanese semiconductor industry. It has banned not only foreign subsidiaries but even joint ventures where the foreign partner retains a controlling interest. At the same time, it has avoided depriving the country of crucial technology from abroad by encouraging licensing and technical assistance agreements between Japanese and foreign firms. The latter, though barred from setting up subsidiaries, have generally been willing to enter into such agreements, and the reluctant exceptions have been enticed to do so by the government's permission to establish joint ventures with Japanese firms.

The government has also had an impact on the fortunes of domestic firms by its influence over licenses, R&D, industry rationalization, and competition. But public policies in these areas apparently have not hindered the development of new firms. This suggests the need to look elsewhere for the major reasons why the receiving tube firms have successfully maintained their large market shares and remained the principal diffusers of semiconductor technology in Japan.

Nature of the Market

The Japanese electronics industry has several distinguishing characteristics: it produces primarily consumer goods and almost no military equipment; it has grown very rapidly over the last two decades from a small initial base; and it sells a large portion of its output in foreign markets, particularly the United States. Since the demand for semiconductors is derived from the demand for final electronic equipment, these features have repercussions in the semiconductor market as well. This section examines how they have affected the opportunities and incentives of the new firms and the receiving tube firms to introduce and use new technology.

Market Composition

The Japanese defense budget has been modest since World War II, and much of the military equipment the country has acquired has come from abroad.[24] Consequently, the portion of electronic output destined for the military is very small, even compared to Europe. By 1960, ten years after the Japanese defense forces were reestablished, annual procurement of military electronic equipment amounted to less than $10 million, or 2 percent of the production of final electronic goods.[25] The relative importance of the military market has not appreciably increased since.

With no significant military market, Japanese producers have concentrated on the commercial market, particularly radios, television sets, and other consumer products. Figure 6-1 illustrates this traditional emphasis, and also shows that in recent years the industrial market has become more important as the production of electronic calculators and computers has grown.

The last chapter argued that the small size of the military market in Europe has hindered the entry and growth of new indigenous firms. Since established producers benefit from considerable learning economies, new firms cannot compete except by concentrating on new devices and processes. Most new semiconductor developments are first used in military products; only after learning economies have pushed costs down, do they penetrate the industrial and consumer markets. So by the time sufficient demand arises in Europe to support domestic production, American firms can provide the new devices at prices below the costs of European producers. New European firms, as a result, seldom have the opportunity to pioneer technology and grow rapidly while exploiting a temporary monopoly based on this technology.

24. This is particularly the case with electronic equipment. When, for example, Japan began producing F-104J aircraft in the early 1960s, nearly all the electronic equipment was imported from the United States. This was necessary because military equipment requires much more sophisticated technology than the commercial products on which the burgeoning Japanese industry has concentrated. See G. R. Hall and R. E. Johnson, "Transfers of United States Aerospace Technology to Japan," in Raymond Vernon (ed.), *The Technology Factor in International Trade,* A Conference of the Universities–National Bureau Committee for Economic Research (New York: Columbia University Press for the National Bureau of Economic Research, 1970), pp. 334–36.

25. Leary, "Electronics in Japan," p. 74.

Figure 6-1. Value of Japanese Electronics Production, by Market, 1956–68[a]

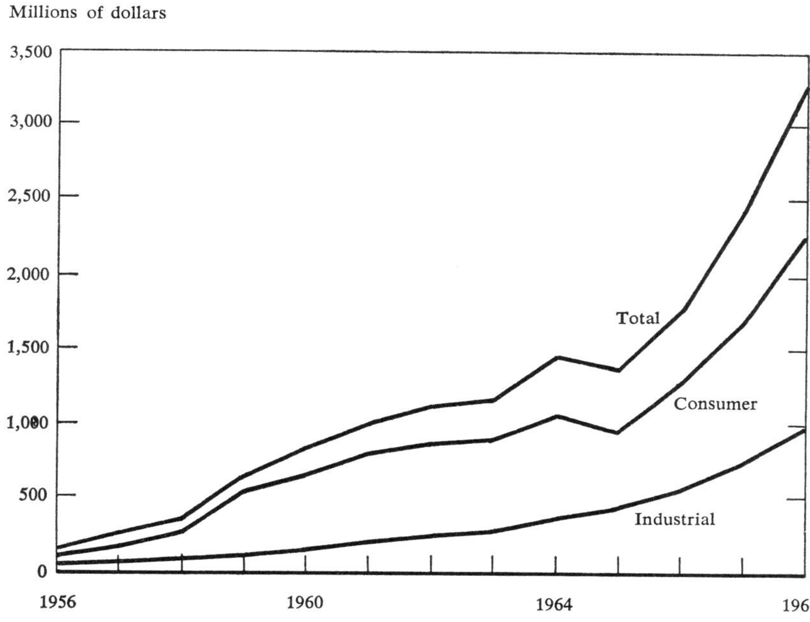

Source: U.S. Department of Commerce, Business and Defense Services Administration, "Japanese Communications and Electronics Production" (periodic reports; processed). These reports are based on data obtained from the Japanese Ministry of International Trade and Industry and the Communications Industries Association of Japan.

a. The small amount of military electronic equipment, which is not separately specified, is included with the industrial market.

The lack of a military market has similarly impeded the entry and growth of new firms in the Japanese industry, even though the government has barred foreign subsidiaries and is usually willing to restrict imports that hinder the expansion of Japanese firms into important new areas. There are two reasons for this. First, since a substantial portion of Japanese final electronic equipment is exported, Japanese semiconductor firms, though protected in their home market, must compete at least indirectly with foreign producers. Otherwise, the overseas market for Japanese electronic goods would suffer, causing the Japanese demand for semiconductors to wither. Second, there are few opportunities for new firms to reap large profits by exploiting in the Japanese market new technology developed by foreign firms since the government normally authorizes licenses for several Japanese firms in any particular technology to promote competition among Japanese producers.

Rapid Growth from a Small Base

The electronic industry has grown rapidly in Japan, more rapidly than in Europe and the United States. As Figure 6-1 illustrates, output of final electronic products increased nineteenfold in Japan in the 1956–68 period, while there was only a fourfold increase in the United States during the same period.[26]

This exceptionally rapid rate of expansion has been possible in part because the Japanese industry started from a very small base. During World War II, the industry was cut off from the rapid advances in technology occurring abroad, and skimped on its own R&D in order to devote more resources to the war effort. In the postwar period, production was curtailed and then redirected. So by the early 1950s, the industry was just beginning to recover from the adverse effects of the war, and its output was still quite modest.

The small size of the industry meant that receiving tube producers had less invested and less to lose in receiving tube production than their American and European counterparts at the time semiconductor technology was germinating. In 1954, for example, all the Japanese producers together accounted for only $15 million worth of receiving tubes.[27] The rapid growth experienced since by the Japanese electronics industry has also helped create a receptive attitude toward change on the part of the receiving tube producers by reducing the risks associated with new products and new technologies and by increasing costs, in terms of declining market shares, to firms content simply to maintain the status quo.

The importance of these factors in stimulating the receiving tube producers to adopt and diffuse new semiconductor technology should not be exaggerated. The American and European electronics markets have

26. United States production is reported in *Electronic Industries Yearbook, 1969* (Washington: Electronic Industries Association, 1969), Table 1. For data on European production which indicate that the electronic industry has grown faster in Japan than Europe, see National Economic Development Office, Electronics Economic Development Committee, *Statistics of the Electronics Industry* (London: NEDO, June 1965), Table 35; and "L'Electronique en France: Statistiques et Réglementation" (Paris: Fédération National des Industries Electroniques, quarterly; processed).

27. In the same year, U.S. production was $300 million. See Japanese Ministry of International Trade and Industry, Year Book of Machinery Statistics (Tokyo, annual; in Japanese; translated for the author); and U.S. Department of Commerce, Business and Defense Services Administration, *Electronic Components: Production and Related Data, 1952–1959* (1960), p. 7.

grown rapidly too, although not as fast as the Japanese market, and yet the receiving tube firms in these countries are not the most aggressive diffusers of new semiconductor technology. Nevertheless, the rapid growth of the Japanese industry from a small postwar beginning has certainly reenforced the proclivity of the receiving tube firms to imitate and exploit new semiconductor developments in Japan.

Trade and Foreign Markets

Foreign markets provide an important outlet for many Japanese electronics products, which contain an important share of the country's semiconductor production. Traditionally, the transistor radio has been among Japan's most important electronic exports. Table 6-5 shows that more than two-thirds of the transistor radios made in Japan have been shipped to the United States and other foreign markets. The table also indicates the importance of this one product for the electronics industry in general and the semiconductor industry in particular. Since the late 1950s, it has accounted for between 11 and 21 percent of the Japanese production of final electronic goods and absorbed between 20 and 67 percent of the country's transistor output.

The prominence of the transistor radio has declined in recent years as the output of tape recorders, computers, and other electronic equipment has expanded. Since a substantial number of these newer products are also exported, Japanese interest in foreign markets has not waned. Their importance compels Japanese semiconductor producers to follow closely new foreign developments in the semiconductor field and to adopt them as soon as economic conditions warrant. These firms are exposed to the stimulation and discipline of foreign competition, even through direct semiconductor exports or imports seldom exceed 15 percent of Japanese output (see Figure 3-5), since a large portion of semiconductor production is used in final electronic products destined for export markets. Should Japanese semiconductor firms fail to remain competitive with foreign producers, exports of final electronic equipment would be adversely affected[28] and the domestic demand for semiconductors reduced. Since the receiving tube firms are both the major semiconductor producers and large manufacturers of final electronic equipment, their own export market for the latter is directly at stake.

28. Unless, of course, final equipment manufacturers switched to foreign semiconductors, but this would still curtail domestic semiconductor production.

Table 6-5. Japanese Transistor Radio Production, Exports, Share of Final

	Transistor radios			
Year	*Production (millions of dollars)*	*Total exports (millions of dollars)*	*Exports to United States (millions of dollars)*	*Total exports as percentage of production*
1957	n.a.	n.a.	n.a.	n.a.
1958	n.a.	n.a.	n.a.	n.a.
1959	122	94	57	*77*
1960	169	119	55	*70*
1961	177	118	48	*67*
1962	189	144	71	*76*
1963	195	157	71	*81*
1964	261	181	74	*69*
1965	236	177	80	*75*
1966	265	227	104	*86*
1967	322	267	120	*83*
1968	357	322	160	*90*

Sources: U.S. Department of Commerce, Business and Defense Services Administration, "Japanese Com-
istry of International Trade and Industry; and Japanese Ministry of Finance, *Japan Exports and Imports:*
a. Assumes each radio has six transistors.
n.a. not available.

From the discussion of the diffusion process and its effect on a country's imports and exports in Chapter 3, it is apparent that Japan's comparative advantage in producing semiconductors depends, first, on the level of its technology, and second, on the price of its production factors compared with those of other countries. Since semiconductor fabrication is labor intensive, this second consideration is governed mainly by wage rates.[29]

When Japanese firms became active in the semiconductor field in the early 1950s and began producing transistor radios, their major competitors were American firms and the American market was the principal battlefield. At that time, Japanese wage rates were well below those of the United States, so the Japanese were able to compete despite superior American technology.[30] Since the 1950s, however, two developments

29. Wage rates, in turn, are affected by exchange rates. By maintaining an undervalued exchange rate during most of the postwar period, Japan has reduced its wage rates (and other factor prices) relative to those of other countries. This has stimulated export-led growth, particularly in activities such as semiconductor production that benefit greatly from learning economies.

30. In final electronic products like computers or missiles where sophisticated

Equipment Output, and Use of Transistors, 1957–68

Final electronic equipment		*Transistor production*		
Total production (millons of dollars)	*Transistor radios as percentage of total production*	*Total (millions of units)*	*Used in radios*[a] *(millions of units)*	*Transistors used in radios as percentage of total*
248	n.a.	6	4	*67*
351	n.a.	27	18	*67*
638	*19*	87	48	*55*
813	*21*	140	67	*48*
1,000	*18*	180	73	*41*
1,115	*17*	232	80	*34*
1,169	*17*	268	94	*35*
1,436	*18*	416	139	*33*
1,360	*17*	454	138	*30*
1,788	*15*	617	159	*26*
2,416	*13*	766	177	*23*
3,258	*11*	939	190	*20*

munications and Electronics Production," periodic reports based on data obtained from the Japanese Min-
Commodity by Country (Japanese Tariff Association, monthly, with December issue giving annual data).

have diminished the advantage Japanese producers derive from low wage rates.

The first is the swift rise in Japanese wages that accompanied the rapid growth of the country's economy during the 1950s and 1960s. Between 1955 and 1967, wages increased 128 percent in the electrical machinery sector; the increase in the United States was only 51 percent.[31]

Second and probably more important, Hong Kong, Taiwan, Korea, Singapore, and other countries with labor rates well below the Japanese began producing semiconductor devices in the 1960s. The diffusion of semiconductor technology to these countries was largely effected by American firms struggling to reduce costs. Faced with growing Japanese competition for the American consumer market and denied access to the relatively cheap Japanese labor, a number of American firms established

semiconductor technology was relatively more important and cost considerations less important, American firms had no problem maintaining their comparative advantage.

31. Japanese wages include bonuses and family allowances, which are important components of labor compensation in Japan. The sampling base for the Japanese data was changed several times over the 1955–67 period. See International Labour Office, *Year Book of Labour Statistics* (Geneva: ILO, annual).

foreign subsidiaries in other countries with even lower wages. Fairchild, for example, set up a semiconductor plant in Hong Kong in 1963 and another in Korea several years later. Signetics and Motorola now have semiconductor operations in Korea too. Philco-Ford has a plant in Taiwan, and Sprague Electric one in Hong Kong.

Rising wage rates in Japan and the diffusion of semiconductor technology to other countries with lower wage rates tend to undermine Japan's comparative advantage in semiconductor production. As a result, Japanese producers have been under considerable pressure to advance the level of their technology and close the gap with the United States.

This pressure and the attitudes and incentives generated by the rapid growth of the Japanese electronic industry have stimulated the receiving tube firms to adopt new semiconductor technology swiftly. They have countered any tendency toward lethargy brought on by the lack of aggressive competition from foreign subsidiaries and new indigenous firms.

Conclusions

Two types of semiconductor producers exist in Japan—receiving tube firms and new firms. Though joint ventures between Japanese and foreign firms do exist, no semiconductor firms in Japan are wholly owned or even controlled by foreign interests.

Like their American and European counterparts, Japanese receiving tube firms are large diversified enterprises whose receiving tube production is but one of many activities. They were among the early manufacturers of semiconductors in Japan. Several new firms were also early producers, but most entered the industry during the latter half of the fifties or during the sixties.

Semiconductor R&D expenditures, at least until recently, have been relatively modest and directed primarily at absorbing foreign technology and adding minor though often ingenious improvements. The only major Japanese innovation was the tunnel diode developed by Sony, a new company that also holds most of the early semiconductor patents acquired by Japanese firms. Over the entire period, however, the receiving tube firms have generated many more patents than the new firms. This reflects the increase in their R&D efforts in the sixties and suggests that they have made a greater contribution to the innovative process than the new firms.

The receiving tube firms have also been the major diffusers of semiconductor technology in Japan. They have captured and held the largest market shares in the industry. In addition, since the end of the fifties, they have introduced most of the new semiconductor devices into Japan and thereby taken the initiative from the new firms in the intercountry transfer of new advances.

The receiving tube producers are the major diffusers of semiconductor technology in Japan for a number of reasons. One important factor is the government ban on foreign subsidiaries and joint ventures controlled by foreign firms. Japanese firms have not had to battle American subsidiaries for their own market as European companies have. Moreover, most foreign firms have willingly provided licenses and technical assistance to Japanese firms. The hesitant exceptions have been pressured to do so in return for permission to set up a joint venture.

The impact of the government on the Japanese semiconductor industry goes far beyond its control over direct foreign investments. Its policies influence licenses and know-how agreements between Japanese and foreign firms, semiconductor R&D effort, industry rationalization, and competition. But government policies do not seem to impair the ability of new firms to enter the semiconductor industry and compete with the receiving tube firms. On the contrary, in some respects they aid new firms.

The primary problem faced by new Japanese firms is a dearth of opportunities. Because of learning economies, new firms can compete with the established firms only by pioneering technology. Since most new developments are first used in military equipment, the opportunities to pioneer are far fewer in Japan with its negligible military market than in the United States.

While the composition of the Japanese semiconductor market has curtailed competition from new indigenous firms and restrictions on direct foreign investment have barred competition from foreign subsidiaries, other factors have stimulated the receiving tube firms to introduce and use new semiconductor techniques quickly. Specifically, the rapid growth of the Japanese electronic industry has increased the cost associated with maintaining the status quo and created an environment favorable to change. The small size of the industry in the early postwar period and its unsettled past have lessened the stultifying influence of vested interests. And rapidly rising Japanese wages, plus the diffusion of semiconductor technology to Hong Kong, Taiwan, and other low-wage countries, have forced the Japanese receiving tube firms to employ new semiconductor technology quickly in order to maintain their important markets abroad.

CHAPTER SEVEN

Conclusions

THE SEMICONDUCTOR INDUSTRY, in its infancy only a generation ago, has grown with remarkable speed. Its rapidly advancing technology has greatly increased the variety of semiconductor devices available, improved their performance capabilities, and reduced their costs. Most of this new technology was developed and introduced by a handful of American companies, and then quickly diffused in the United States, Europe, and Japan.

This final chapter draws together and evaluates the evidence presented in earlier chapters concerning the importance of new firms and easy entry conditions for rapid diffusion. This leads to an examination of the problems encountered by imitating countries striving to catch up with countries using more advanced technology and at the same time to be pioneers in diffusing at least some new techniques. Finally, some general conclusions based on the findings for the semiconductor industry are considered, along with their implications for public policy.

New Firms and Diffusion

Chapter 1 advanced the proposition that the diffusion of new technology is stimulated by a flexible market structure that allows new firms to arise and replace the established industry leaders whenever the latter delay too long in using new techniques. The development of the American semiconductor industry strongly supports this hypothesis. The development of the European and Japanese industries, although not conflicting with the hypothesis, does suggest certain modifications and extensions.

Diffusion in the United States

New semiconductor technology has diffused faster in the United States than in any other country. This is apparent from the country's very short imitation lag, from the rapid growth of semiconductors both in absolute terms and relative to its total active-component production, and from the large proportion of integrated circuits and other sophisticated devices in its semiconductor output.

Although the large American electrical companies engaged in the production of receiving tubes were aware that semiconductors posed a threat to their tube operations and were consequently among the earliest producers of semiconductors, new firms with little or no previous experience in the active-components industry have been the most aggressive diffusers of new semiconductor technology. They have been particularly adept at taking important new developments from the laboratory, introducing them into high-volume production lines, and achieving acceptable yields or rejection rates. The dominant market shares captured by Texas Instruments and several other new firms are evidence of the success of these firms in this difficult task. New firms with more modest market shares have also encouraged the swift diffusion of new technology in certain subsectors of the semiconductor industry. General Instrument and American Micro-Systems, for example, stimulated the adoption of MOS techniques. In addition, the mere fact that new firms can enter the industry and exploit new developments neglected by established firms helps motivate the latter to remain alert for new technological possibilities.

Often we think of a new firm as a small newly established concern, but other types of new companies have also contributed to diffusion in the American semiconductor industry. Many firms were active in other industries when they expanded their operations to include semiconductors, and some were not small. Very few of the latter, however, became major diffusers of semiconductor technology; Motorola is the principal exception.

The hypothesis linking new firms and easy entry conditions with rapid diffusion is impressively supported by the American example, because the new firms have been rapid diffusers even though they have not been major innovators of new semiconductor technology and hence not in the innovator's favorable position for exploiting its new technology in the marketplace. Rather, most of the major process and product innovations have come from Bell Laboratories and the receiving tube firms. These firms have also spent more heavily (both in absolute terms and as a percentage of semiconductor sales) on semiconductor research and development (R&D) and have acquired the greatest number of semiconductor patents.

Why have Bell Laboratories and the receiving tube firms contributed more to the creation of semiconductor technology but less to its diffusion than new firms? The answer involves differences between types of firms in capabilities and incentives. Bell Laboratories and the receiving tube firms are large concerns with command of substantial financial resources. They can afford large R&D expenditures and in particular support basic research with its longer and much more uncertain payoff period. A number of the major advances in semiconductor technology (for example, the first transistor) arose from such fundamental research. These firms were strongly motivated to undertake semiconductor R&D as a form of insurance against being closed out of the new industry by vigorous newcomers. And important patents could be traded with firms both at home and abroad for other technology.

The new firms, on the other hand, had more incentives to exploit new semiconductor technology quickly. First, they had no vested interests in receiving tube activities. Second, unlike Bell Laboratories and the receiving tube firms, most new firms had no past experience with antitrust litigation and hence no reservations about trying to dominate the industry. Third, small new firms had to pioneer the use of new technology, for this was the only way to get around the entry barrier erected by learning economies. The receiving tube firms were large enough to absorb the costs imposed on latecomers and could delay without fear of being barred forever from the market. Fourth, the semiconductor industry offered the small new firm the possibility of very rapid growth with substantial personal rewards for managers and owners. In the receiving tube firms, semiconductor sales, particularly in the early years of production, could not greatly affect overall growth because these firms were already so large and diversified.

In addition to being more motivated, new firms apparently are more adept at diffusion. Their smaller size permits more rapid and effective communication within the firm, allowing a faster response to changing market opportunities. And they can attract the type of person willing to accept the great risks associated with pioneering the use of new technology since they can provide rewards commensurate with the risks.

These incentives and capabilities, which encourage new firms to be aggressive diffusers of new technology, depend on two attributes. The first is newness, but only in the sense that a company is new to the industry and thus has no vested interests to protect, and not in the more restrictive sense of being newly established. The second is size. This explains why the most successful diffusers are newly created companies and small new companies established in other industries, but rarely large new firms from other industries.

New firms have been able to enter the semiconductor industry and make their contribution to the rapid diffusion of new technology because the industry has followed a liberal licensing policy and tolerated high interfirm mobility of scientists and other professional personnel. The insignificance of scale economies has also been important in keeping entry barriers low, as have certain government activities. The Defense Department, for example, has not hesitated to authorize the procurement of semiconductors from new firms for military equipment. This not only opens up the sizable military market to new firms, but also helps them gain acceptance in the commercial market. Unlike its policy on procurement, the government does discriminate in awarding R&D contracts in favor of established firms. But the adverse effect this has on entry is largely mitigated by the liberal licensing policy and the great mobility of scientists and engineers that arose in this industry partly because of government antitrust activities.

Diffusion in Europe and Japan

New semiconductor technology, though developed largely in the United States, has diffused very quickly in Britain, France, Germany, and Japan. These countries have produced most of the major semiconductor devices two to three years after their introduction in the United States. Similarly, the growth of semiconductor output as a percentage of active-component production has seldom lagged far behind that of the United States. Indeed, the differences among countries in the rate of diffusion are so small that they may well be entirely warranted by differences in the size and composition of semiconductor markets. The large American market, with its great demand for new and sophisticated military devices, can support the domestic production of most new semiconductor devices several years before the European and Japanese markets.

The rapid pace of diffusion in Europe and Japan cannot be attributed to new firms as in the United States. New firms have sprung up in the semiconductor industries of Europe and Japan, but with few exceptions they have remained small, specialized producers. To a great extent, these firms are blocked by learning economies from challenging the established firms in the important large-volume markets. They cannot skirt this barrier by pioneering the use of new technology, because in size and composition their markets are unlike the American market, which greatly favors this activity. By the time sufficient demand arises in Europe and Japan to support local production, new firms are already at a competitive disadvantage with American subsidiaries and large established firms that can afford licensing and technical assistance agreements with American firms.

Thus the market structure of the semiconductor industry in Europe and Japan differs markedly from that in the United States. In the latter country, entry barriers have been low, many new firms have entered the industry, and a few have risen to become industry leaders. In Europe and Japan, entry barriers have been high. Although some new firms have entered the industry, they are not numerous, they have not become industry leaders, and they have not played an important role in diffusing new technology. Why, then, has diffusion proceeded so rapidly in these countries?

In Europe, foreign subsidiaries have assured the swift diffusion of new technology when the established industry leaders have faltered. More specifically, in the 1960s, when the European receiving tube firms, which, like their American counterparts, are large, diversified companies, delayed too long in providing devices made with the planar silicon technology, American semiconductor firms set up subsidiaries in Europe and concentrated on the new devices neglected by the industry leaders. As a result, they captured a substantial share of the European market which, in turn, stimulated the European receiving tube firms to adopt the new technology. Thus, foreign subsidiaries in Europe, like new firms in the United States, have provided an alternative to the established semiconductor leaders and imposed a discipline on the industry that assures the swift diffusion of new technology.

It is interesting that the foreign subsidiaries instrumental in the diffusion of new technology in Europe are the *new* foreign subsidiaries set up since the war and mostly in the 1960s. The older foreign subsidiaries, such as International Telephone and Telegraph (ITT), have behaved much more like the established European firms. This suggests that easy entry conditions and a flexible market structure do play an important role in stimulating diffusion in Europe, but that these favorable market structure characteristics extend only to a certain type of new firm, namely, foreign subsidiaries. These firms manage to evade the entry barriers blocking new domestic firms because their parent companies share with them much of the know-how derived from past production experience, provide specialists and other professional personnel, and if necessary supply venture capital.

In Japan, the receiving tube firms, most of which again are large diversified concerns, soon became the major semiconductor producers and have managed to remain leaders in the industry, partly because high entry barriers have stifled new firms and barred foreign subsidiaries. Like their European counterparts, new domestic firms are hampered by learning economies. New foreign subsidiaries could avoid this obstacle, but they run into an equally formidable one—a government ban on firms controlled by foreign interests.

Although the Japanese receiving tube firms have lacked competitive pressures from either new firms or foreign subsidiaries, they have still introduced new semiconductor technology quickly. There are several reasons for this.

First, the receiving tube firms have been subjected to other forms of discipline. Since a large portion of Japanese semiconductor output is destined for foreign markets, either directly or after incorporation in final electronic equipment, the established semiconductor producers must remain competitive with the best firms in the United States and elsewhere or yield their important foreign markets. This means that they must follow closely the new semiconductor technology arising abroad and employ it quickly once their economic conditions warrant its use. The pressure for speedy diffusion has increased as the comparative advantage Japan once enjoyed from its low wage rates has been eroded by rapidly rising Japanese wages and the transfer of semiconductor technology to Hong Kong, Taiwan, and other countries where wages are substantially below those in Japan. Supplementing the stimulating effects of foreign competition is the considerable competition fostered by the government among the receiving tube firms. It has, for example, refused to authorize foreign licensing agreements that grant any of these companies the exclusive use of semiconductor technology in Japan.

Second, the small size of the Japanese electronic industry in the early postwar period, its unsettled past, and its very rapid growth since have probably reduced the Japanese receiving tube firms' tendency to maintain the status quo and increased their receptivity to new technology.

Third, American firms have generally been very cooperative. With few exceptions, they have been willing to license Japanese as well as other foreign firms and aid them in assimilating new semiconductor technology, even though in the process they are helping establish potential rivals.

The willingness of American firms to facilitate the transfer of semiconductor technology to foreign firms and thereby propagate competitors abroad can be attributed in part to the market structure that developed in the American industry. Rarely has any one firm supplied over 25 percent of the American market, and for most firms market shares are quite small. Consequently, when one firm licenses another and the latter manages to expand its sales as a result of the new technology acquired, only a small fraction of this increase comes at the expense of the licenser—most is instead at the expense of its competitors. In addition, the royalties received can be reinvested in R&D, helping the firm maintain its technology and, in turn, its share of the market. Since a number of firms possess advanced semiconductor technology, a company that refuses to provide technical assistance may merely force a prospective licensee to another firm, thereby diverting royalties from itself to a competitor without preventing the dissemination of the coveted technology.

If one firm dominated the American semiconductor industry (as is the case, for example, in the computer and photocopying industries), a prospective licensee would have few alternatives if turned down by the major firm. Furthermore, the output of licensees would be largely, perhaps entirely, at the expense of the firm being asked to provide licenses and technical assistance. Under such conditions, it is unlikely that American technology would be so willingly provided to foreign firms.

The evidence presented in this study about the market structure characteristics conducive to the rapid diffusion of new technology shows that easy entry conditions and new firms were very instrumental in the rapid utilization of new semiconductor techniques and devices in the American market. For a number of reasons, these firms are highly motivated to pioneer the use of new developments, and often are more successful in exploiting new technology in the marketplace than the large established firms responsible for most of the innovative efforts. In addition, the presence of a number of major American semiconductor firms and the highly competitive climate in which they operate have stimulated the transfer of American technology to foreign firms.

The European experience identifies a different type of new firm that can assure rapid diffusion—the new foreign subsidiary. With the entry of new indigenous firms impeded, new foreign subsidiaries have become important diffusers of new technology, providing an alternative to the industry leaders when they delay too long in using new technology.

Japan demonstrates that the large established firms, under certain conditions, will employ new technology quickly. Specifically, very rapid growth and dependence on foreign markets have exposed the established firms to competition from each other and from foreign firms and eliminated the need for the alternative diffusion channels that new firms and foreign subsidiaries provide.

A Dilemma for Imitating Countries

This study indicates that imitating countries may face a dilemma in determining the optimum market structure for diffusion if they are concerned both with catching up with the innovating country in the use of new technology and with pioneering the diffusion of at least some new technology. The latter involves surpassing the innovating country in the use of certain technology. American experience indicates that new firms are particularly good at this last function, while the experience of Europe and Japan suggests that large established firms are needed to catch up unless a country is willing to become dependent on foreign subsidiaries.

Although easy entry and high concentration are not necessarily mutually exclusive,[1] in practice the same forces that generate a market structure dominated by a few large firms often impede entry as well. Imitating countries trying to rationalize their semiconductor industries run into this problem. Britain and France, for example, award very few government R&D and production contracts to new firms trying to enter the semiconductor industry, for this would divert support from the few large firms chosen as government instruments for developing viable alternatives to American subsidiaries. This policy raises the barriers to entry in these countries and probably reduces their propensity to pioneer the use of new semiconductor technology.

This does not necessarily mean, however, that the present rationaliza-

1. Conceivably, for example, public policy could pressure small lethargic firms to leave the industry or merge with larger firms and at the same time reduce the entry barriers to new aggressive firms.

tion policies under way in Britain and France are misguided, for this may be the only way these countries can catch up and keep pace with American technology without completely turning over their semiconductor industries to foreign interests. Moreover, the short-term costs of rationalization are probably not great, since even if the barriers to entry arising from such government policies were completely eliminated, the major barrier erected by learning economies would remain.

Still, in creating a few large semiconductor firms, public policy should strive to minimize the adverse effects on entry conditions and market structure flexibility. This would increase the probability that when opportunities do arise for the country to pioneer the use of new technology they will not be lost. Even now, occasional opportunities may occur, and as the semiconductor industry matures, the market conditions that give the United States an advantage in innovating and pioneering the use of new technology may change. Should this happen, Europe and Japan may need new firms to seize and exploit the resulting opportunities and thereby become world leaders, not just rapid followers, in the diffusion of semiconductor technology.

Generalizations and Public Policy

It is always hazardous to generalize on the basis of one observation or, as in this case, on the basis of one industry study. Consequently, most of this dangerous task will be left to the reader. By way of assistance, however, this final section looks at certain important characteristics of semiconductor technology and suggests that they are not as unique as they may at first appear. It then examines a few areas where generalizations seem feasible.

Semiconductor Technology

The absence of large economies of scale and the presence of sizable learning economies are two of the most important characteristics of semiconductor technology affecting the market structure of this industry and the types of firms responsible for diffusion around the world. The first has made it possible for small new firms to enter the industry and contribute to the diffusion of new technology. The second has forced such firms to concentrate on pioneering the use of new technology, and also has helped determine the shifting military-industrial-consumer market pattern that prevails. It is this pattern, coupled with differences in the size and composition of semiconductor markets, that (1) gives American firms an advantage in innovating technology and pioneering its diffusion, (2) creates a need in other countries for large producers to effect the swift imitation of American technology, and (3) stifles the opportunities in other countries for new firms (aside from new foreign subsidiaries) to arise and challenge the established industry leaders.

The importance of negligible scale economies and significant learning economies in shaping diffusion in the semiconductor industry raises the question of how typical such conditions are in other industries. A complete answer lies outside the scope of this study, but it does appear that these features frequently characterize industries in the early stages of technological development—the rapid-growth research-intensive industries.[2] For at this stage of an industry's development, its technology is advancing most rapidly and much can be learned in the process of production. Moreover, firms are often reluctant to invest heavily in equipment that may soon be obsolete. There are, of course, exceptions—new industries where substantial economies of scale are possible in R&D, production, or distribution right from the first. But generally such economies arise as an industry matures.[3]

Some General Conclusions and Implications for Public Policy

The semiconductor industry indicates that the type of firm contributing most to *technological progress* may vary over the technological development cycle: the type of firm most responsible for the creation of new

2. It has even been suggested that nearly all industries benefit from learning economies (defined to include all cost reductions associated with increased cumulative output), and that their effect on costs in stagnant or slow-growing industries is often obscured by inflation and other extraneous factors. See Boston Consulting Group, *Perspectives on Experience* (Boston: BCG, 1970).

3. See Raymond Vernon, "International Investment and International Trade in the Product Cycle," *Quarterly Journal of Economics,* Vol. 80 (May 1966), pp. 190–207; Seev Hirsch, *Location of Industry and International Competitiveness* (London: Oxford University Press, 1967); and Dennis C. Mueller and John E. Tilton, "Research and Development Costs as a Barrier to Entry," *Canadian Journal of Economics,* Vol. 2 (November 1969), pp. 570–79.

technology[4] may differ from the type most responsible for its diffusion. Although the latter generally reaps the greatest financial rewards, both types have contributed to technological progress.

As a result, tests of the Schumpeterian and related hypotheses that attempt to identify the type of firm most responsible for progress are often misleading. Many relate company characteristics (such as size) to company contributions toward technological progress, as measured by the number of patents acquired, the number of major innovations introduced, or some similarly limited measure that identifies company contributions only at one point or along one segment of the technological development cycle. Public policy, however, should be concerned with stimulating progress, and with the firms responsible, over the entire cycle.

Easy entry conditions help assure speedy intracountry diffusion of new technology, for they permit new firms to arise and supplant the established industry leaders whenever the latter delay too long in introducing new products and processes. New firms are particularly effective at disciplining the established firms because they are highly motivated to exploit new technology in the marketplace just as soon as its use is warranted.

Easy entry conditions in the innovating country, which tend to keep the level of concentration low, also stimulate intercountry diffusion, because firms with many competitors and small market shares are generally more willing to provide licenses and technical assistance to foreign firms. As a result, the market structure of the innovating country is a matter of some concern to imitating countries, even though they have little or no influence in shaping it.

Public policy can foster easy entry conditions in a number of ways. Antitrust or competition policies, for example, can prevent restrictive patent pools and assure all firms access to crucial technology on reasonable terms. Government R&D support can also help new firms get started in the research-intensive industries. But the beneficial effects are reduced to the extent that the government, in awarding such funds, discriminates in favor of established firms, concentrates on government laboratories and universities (rather than new firms), and prohibits use of the funds for production-improvement projects.

4. It is even possible that the importance of company types varies in different stages of the innovative process. The type of firm making the breakthroughs at the basic research level may not be the same type that follows up these breakthroughs with applied research and development and eventually introduces new products or processes in the marketplace.

Another important government activity affecting entry is procurement—production contracts can be more important in helping new firms enter an industry than government R&D support. Such contracts not only provide desirable business, but help firms gain acceptability in commercial markets. In addition, the military market is important for new firms trying to introduce new products that initially are too expensive for the industrial or consumer market.

Because of *differences in demand conditions* arising from variations in the size and composition of national markets, certain countries enjoy a comparative advantage in the introduction and initial diffusion of new technology in some industries. Other countries with less favorable demand conditions should strive to imitate the innovating country as quickly as using the new technology becomes warranted. Where learning economies accrue to early producers, as in a number of the new research-intensive industries, small new firms will not generally be effective instruments for this catching-up process. Instead, large firms are needed that can pay for licenses and technical assistance and absorb losses while acquiring experience and know-how during the early stages of production.

Thus rationalization programs to create one or a few large firms may be appropriate in imitating countries and necessary to assure the rapid imitation of technology already in use elsewhere. However, such programs should be designed to minimize their adverse effects on entry conditions so that when opportunities do arise to pioneer the use of new technology they are not lost.

New foreign subsidiaries provide an alternative to large established firms for quickly transferring new technology from the innovating to the imitating countries. These firms are not hampered by learning economies as are new domestic firms, because they have special access to the technology and know-how of those firms in the innovating country that benefit from such economies. Moreover, as a result of their special ties, they can often acquire new technology from abroad as fast or faster than the large independent firms, even though the latter may have licenses and technical assistance agreements with the major foreign producers.

Despite such advantages, there is considerable reluctance on the part of many countries to allow firms controlled by foreign interests to dominate the strategic and prestigious research-intensive industries. This, they feel, undermines national sovereignty. Moreover, it is stated in a modern version of the infant-industry argument, domestic firms will never become competitive unless they have the opportunity to acquire production experience and, in turn, the benefits of learning economies. For these reasons, some imitating countries have limited the activities of foreign subsidiaries.

This study suggests that it is possible to ban foreign-controlled firms and still quickly acquire new technology from abroad; at least, Japan has been able to do so in the semiconductor industry. However, restricting foreign subsidiaries entails certain risks. Such firms, as the European experience in semiconductors demonstrates, provide insurance that new technology will quickly disseminate in the imitating country when indigenous firms fail to respond as quickly as desirable.

The risks involved in barring foreign subsidiaries are reduced when the established firms are exposed to other forms of discipline. This may come from new domestic firms if entry barriers are low, from other established firms if competition for the domestic market is vigorous, or from foreign firms if overseas markets are important. While the risks may be reduced, they cannot be completely eliminated, for in barring new foreign subsidiaries, an imitating country cuts off a potential diffusion channel that just may be needed. Of course, this disadvantage may be offset by other considerations. This is a political decision that involves weighing competing national goals.

APPENDIX

Company Size, Spillover, and Government R&D Support

THIS APPENDIX uses regression analysis to analyze the research and development (R&D) efforts of firms in the American semiconductor industry during 1959, the one year for which adequate data for individual firms are available. The purpose of this investigation is to determine if large firms received more government support for R&D in the semiconductor field than did small firms and if this support, which was primarily for R&D on military requirements, produced any spillover of value to firms in their commercial markets.

Thirty-eight firms, responsible for over 95 percent of the American semiconductor output in 1959, are covered in the analysis. They vary in size from large diversified enterprises such as General Motors and General Electric to very small firms whose 1959 sales were under $50 thousand. For each firm, the following data are considered: expenditures on semiconductor R&D during 1959 financed by the government, those in 1959 financed by the firm itself, semiconductor sales in 1959, total sales in 1959, and the average number of semiconductor patents received annually during 1962–64.[1]

Company Size and Government R&D Support

Equation (1) indicates the relation estimated between the amount of government R&D funds that a firm received (G_i) and its semiconductor

1. Government-funded R&D expenditures, company-funded R&D expenditures, and semiconductor sales were obtained from the U.S. Department of Defense, Survey of 64 Semiconductor Companies, 1960, unpublished tabulations. Semiconductor patents were collected at the U.S. Patent Office. Total sales for all but eight firms are from *Moody's Industrial Manual* (1960). All but one of the exceptions are small firms, which presumably concentrated primarily, if not entirely, on semiconductor production. Semiconductor sales were therefore used as an estimate of their total sales. Total sales for the one remaining exception were estimated on the basis of other information. For additional comments on the data, see the notes to Table 4-4.

sales (S_i), with both variables measured in millions of dollars:

$$G_i/S_i = -\underset{(0.0015)}{0.0003} \cdot 1/S_i + \underset{(0.025)}{0.058} \qquad R^2 = 0.00 \tag{1}$$

(The figures in parentheses are standard errors, and R^2 is the coefficient of determination.[2]) To alleviate heteroscedasticity, both sides of the equation are deflated by S_i, the independent variable. Consequently, it is the coefficient of the constant term and its standard error that are of interest. They indicate that the amount of government R&D funds that a firm receives tends to increase significantly (at the 95 percent probability level) with semiconductor sales. Specifically, a $1 million increase in semiconductor sales is associated with an additional $58 thousand in government R&D support.

Though it may hamper entry, greater government R&D support for the large established semiconductor producers is to some extent justified, since these firms generally have a greater capability for carrying on semiconductor research. Moreover, their large sales are in part a tribute to their past R&D performance. If preferential treatment proportional to differences in semiconductor output is justified, the government does not appear to have discriminated unduly in favor of the larger firms. As equations (2) and (3) indicate, the amount of government R&D support (G_i) a firm receives, relative to its semiconductor sales (S_i), does not increase with firm size measured either by semiconductor sales (S_i) or total sales (T_i).

$$G_i/S_i = \underset{(0.028)}{0.064} - \underset{(0.0015)}{0.0007}\, S_i \qquad R^2 = 0.01 \tag{2}$$

$$G_i/S_i = \underset{(0.025)}{0.058} - \underset{(0.012)}{0.002}\, T_i \qquad R^2 = 0.00 \tag{3}$$

The coefficients of S_i and T_i are both negative and not significantly different from zero. Again, data are in millions of dollars with the exception of total sales (T_i) which are in billions of dollars.

2. Because the equation is deflated by S_i, R^2 does not indicate the proportion of interfirm variation in government R&D support explained by semiconductor sales. Rather, it shows the proportion of interfirm variation in government R&D support per dollar of semiconductor sales that is explained by the reciprocal of semiconductor sales, which is of little interest.

Spillover from Government R&D Support

Although the government may not discriminate unduly against small firms, as equation (1) shows, the amount of R&D it finances tends to increase with the semiconductor output of the firm. This helps the larger producers maintain and expand their military sales and, if the spillover from government R&D support is significant, their commercial sales as well.

The importance of spillover is revealed by equation (4), which relates expenditures on semiconductor R&D that were government financed (G_i) and those that were firm financed (F_i), both in millions of dollars, with the average annual number of patents (P_i) that firms received during 1962–64:

$$(4) \qquad \underset{}{P_i} = \underset{(0.58)}{0.59} + \underset{(0.41)}{2.20}\, F_i + \underset{(0.76)}{2.23}\, G_i \qquad R^2 = 0.82$$

According to the equation, no significant difference exists between the number of patents generated by government-funded R&D and the number generated by company-funded R&D.[3] An extra $1 million of R&D expenditures is associated with an additional 2.2 patents whether the expenditures are financed by the firm or by the government.

Since patents acquired as a result of government-funded R&D projects must be licensed royalty free to government contractors and subcontractors when needed for the production of military equipment, there is little incentive to patent new developments unless there is some possibility that they can be used to produce semiconductors for the commercial market. However, in interpreting the equation results, the fact that patents arising from government-funded R&D are less often used commercially than patents arising from privately funded R&D should be taken into account.[4]

It should also be noted that firms when queried have identified very few patents used in producing commercial semiconductors as having

3. This finding does not change when the equation is deflated by the square root of total firm R&D expenditures ($F_i + G_i$) or by the square root of company semiconductor sales to eliminate heteroscedasticity.

4. Mary A. Holman, "The Utilization of Government-Owned Patented Inventions," *Patent, Trademark, and Copyright Journal of Research and Education,* Vol. 7 (Summer and Fall 1963), pp. 109–61 and 321–75.

arisen from government-sponsored R&D projects.[5] This may be explained in part by the reluctance of firms to detract from the R&D they finance, by their failure to report all patents from government-sponsored R&D, and by the less frequent commercial use of patents from government-funded R&D than from privately funded R&D. But it also suggests that the spillover identified by the equation may be indirect rather than direct. That is, military R&D may produce few specific semiconductor devices that firms can sell directly in the commercial market, but instead may generate new know-how and knowledge that firms draw on in conducting their own privately financed R&D efforts to develop better semiconductors for the commercial market. This possibility is also supported by the fact that very few major semiconductor innovations have arisen directly from government-sponsored R&D.[6]

5. *Patent Practices of the Department of Defense,* Preliminary Report of the Subcommittee on Patents, Trademarks, and Copyrights of the Senate Committee on the Judiciary, 87 Cong. 1 sess. (1961), App. D; and John G. Welles and others, "The Commercial Application of Missile/Space Technology," Denver Research Institute, University of Denver (Denver: DRI, 1963; processed), pp. 69–76.

6. See p. 95.

Index